# SpringerBriefs in Applied Sciences and Technology

SpringerBriefs present concise summaries of cutting-edge research and practical applications across a wide spectrum of fields. Featuring compact volumes of 50 to 125 pages, the series covers a range of content from professional to academic.

Typical publications can be:

- A timely report of state-of-the art methods
- An introduction to or a manual for the application of mathematical or computer techniques
- A bridge between new research results, as published in journal articles
- A snapshot of a hot or emerging topic
- An in-depth case study
- A presentation of core concepts that students must understand in order to make independent contributions

SpringerBriefs are characterized by fast, global electronic dissemination, standard publishing contracts, standardized manuscript preparation and formatting guidelines, and expedited production schedules.

On the one hand, **SpringerBriefs in Applied Sciences and Technology** are devoted to the publication of fundamentals and applications within the different classical engineering disciplines as well as in interdisciplinary fields that recently emerged between these areas. On the other hand, as the boundary separating fundamental research and applied technology is more and more dissolving, this series is particularly open to trans-disciplinary topics between fundamental science and engineering.

Indexed by EI-Compendex, SCOPUS and Springerlink.

Rohana Hassan · Azmi Ibrahim · Zakiah Ahmad

# Timber Connections

Mortise and Tenon Structural Design

Rohana Hassan
IIESM
Universiti Teknologi MARA
Shah Alam, Selangor, Malaysia

Azmi Ibrahim
School of Civil Engineering
Universiti Teknologi MARA
Shah Alam, Selangor, Malaysia

Zakiah Ahmad
School of Civil Engineering
Universiti Teknologi MARA
Shah Alam, Selangor, Malaysia

ISSN 2191-530X          ISSN 2191-5318  (electronic)
SpringerBriefs in Applied Sciences and Technology
ISBN 978-981-19-2696-9          ISBN 978-981-19-2697-6  (eBook)
https://doi.org/10.1007/978-981-19-2697-6

This Springer imprint is published by the registered company Springer Nature Singapore Pte Ltd.
The registered company address is: 152 Beach Road, #21-01/04 Gateway East, Singapore 189721,
Singapore

# Preface

With the Grace and Love of ALLAH SWT, this book is published under the title *Timber Connection: Mortise and Tenon Structural Basic Design*. This book was successfully produced with the best level of quality along with the latest facts from the current world of research. God willing, through a mature writing style in elaborating facts and discussions, it can attract readers from all walks of life, whether among students, industry or government. The information contained in this book is very suitable and useful as reference material and guidance, especially for those who are directly involved in the construction and learning industry in the country. Hopefully with a little effort in sharing experience and knowledge available in order to be a guide to people who are interested in optimising timber engineering technology.

Indirectly, the information conveyed through this book can encourage students, researchers, academics, industry and even governments to explore timber technologies and development of connections specifically. The book has seven chapters that discuss the relationship of certain wood characteristics to mortise and tenon in relations.

It begins with Chap. 1 consists of a brief explanation of what is this book aimed for, introduction about timber connections for structural applications and the basic design of mortise and tenon, ended by conclusions. Chapter 2 reviews of current type of mechanical fasteners and their affecting parameters. Chapter 3 describes the structural behaviour of mortise and tenon joints that encounter principle of bending, shear and tensile capacity. Chapter 4 describes the application of mathematical and fundamental theory as stated in the relevant standards, namely EYM, NDS, 2005 and EC 5, 2008. Chapter 5 contains a concise historical review of a laterally loaded connection in timber which leads to the background of the current EYM. The report included related modifications proposed by few researches into the use of wood dowel into the connections. Chapter 6 describes common related parameters that relate to dowel-bearing strength. These parameters of concerns are the dowel diameter, grain directions, moisture contents, specific gravity and density of wood. Chapter 7 describes basic theory on the predictions of the factor of safety.

On this occasion, we would like to express our deepest appreciation and thanks to all parties involved in the successful writing of this book. Thanks for the guidance, giving thoughtful ideas, making the writing of this book the most valuable reference in the development of timber engineering. Mentioned here by, the assistance of additional professor expertise; Prof. Pete Walker from Bath University. Not to be forgotten to all the technicians in the laboratory of the Faculty of Civil Engineering, Universiti Teknologi MARA, Shah Alam for their assistance in preparing the specimens for testing, as well as to emphasise its use. Financial support from grant providers, Universiti Teknologi MARA (600-RMI/DANA 5/3 /RIF (74/2012)) and the Ministry of Higher Education Malaysia (600-RMI/ERGS 5/3 (25/2012) and 600-RMI/RAGS 5/3 (61/2013)).

Hopefully, through reading, research and curiosity about the greatness of science is the key to the success of new technologies that have a foothold for the future. In addition, it is also noted that all technical information and facts stated in this book are a continuation of several research grants. On this occasion, we would also like to thank our beloved family members for always providing support and encouragement in completing the writing of this book. Not to be forgotten is *Springer Nature*, for successfully publishing this book as part of the nation's scholarly assets.

All Praise is Due to Allah.

Shah Alam, Malaysia                                  Associate Professor Ts. Dr. Rohana Hassan
                                                                         Professor Dr. Azmi Ibrahim
                                                                         Professor Dr. Zakiah Ahmad

# Contents

# Symbols

| | |
|---|---|
| $M_A$ | Moment of rotation at A |
| $P_1$ | Value of force at a distance from the mortise face |
| $d_1$ | Perpendicular distance of load to the centre of rotation |
| $F_4$ | The bending moment resistance of the tenon |
| $t$ | Thickness of the tenon |
| $d$ | The depth of the tenon |
| $S_4$ | The modulus of rupture of the material of which the tenon is constructed |
| $k$ | A form factor for round beams, namely 1.18 |
| $D$ | Bolt/tenon diameter |
| $F_e$ | Dowel-bearing strength of wood |
| $F_{e//}$ | Dowel-bearing strength of wood in parallel to grain |
| $F_{e\perp}$ | Dowel-bearing strength of wood in perpendicular to grain |
| $F_{vy}$ | Average mode V yield shear stress |
| $G_{dowel}$ | Specific gravity of wood dowel |
| $G_{base}$ | Specific gravity of wood base/specimen |
| $M_{max}$ | Maximum bending moment |
| $\sigma_{max}$ | Bending stress or bending yield strength |
| $r$ | Radius |
| $F$ | Force resultant used to derive maximum moment |
| $M_{pl}$ | Plastic bending moment |
| $M_{el}$ | Elastic bending moment |
| $G$ | Specific gravity |
| m | Moisture content |
| $w_1$ | Initial weight of specimen |
| $w_0$ | Oven dry weight of specimen |
| $M$ | Bending moment in dowel |
| $S_p$ | Plastic section modulus for dowel |
| $P_y$ | Yield load of a dowel |
| $S_{bp}$ | Spacing of bearing points |
| $\mu$ | Ductility index |
| $\delta u$ | Displacement at fracture |

$\delta y$        Displacement at yield
$P_{0.05}$      5% offset yield load from the load–displacement curve
$f_{(h,0,k)}$     Dowel-bearing strength parallel to grain (EC5)
$\rho$          Density

# Abbreviations

| | |
|---|---|
| AFPA | American Forest & Paper Association |
| ASTM | American Society for Testing and Materials |
| BS | British Standard |
| CFRP | Carbon Fibre Reinforced Polymer |
| CIDB | Construction Industry Development Board Malaysia |
| EC 5, 2008 | Eurocode, 2008 |
| EWP | Engineered Wood Product |
| EYM | European Yield Model |
| *FoS* | Factor of Safety |
| FRP | Fibre Reinforced Polymer |
| GFRP | Glass Fibre Reinforced Polymer |
| GLTs | Glued Laminated Timbers |
| J1, J2, J3, J4 and J5 | Five Joints Group |
| LVDT | Linear Variable Displacement Transducer |
| LVL | Laminated Veneer Lumber |
| MOE | Modulus of Elasticity |
| MS | Malaysian Standard |
| NDS, 2005 | National Design Specifications, 2005 |
| SG | Strength Group |
| TFEC | Timber Frame Engineering Council |

# List of Figures

# List of Tables

# Chapter 1
# Timber Connections for Structural Applications

## 1.1 Introduction

Timber is well known as one of the important materials in the construction industry. It is used in many structural applications around the world due to its good strength-to-weight ratio. Much research has been done on timber alone or as timber composite materials in enhancing its uses as a construction material. In a structure either using timber or timber composites, timber jointing plays a very important role as all timber structures are made of elements that must be fastened or connected together. Hence, a good understanding of the behaviour and the use of the fasteners are necessary in the proper design and construction of timber structures. Timber joints have always been the weak link in timber construction and in the design of large timber structures. The joints being the weakest part of the structure are often the critical points (Fig. 1.1).

Mechanical connections commonly demonstrate a very plastic behaviour due to the dowel-bearing strength of timber itself and the ability of the timber elements to dissipate energy (Stehn and Johansson 2002). Natterer and Sanders (1995) are the successful scientist who studied about the timber design and connections. They reported that in order to produce and optimise a design, few main criteria need to be considered.

These criteria are such as the location, orientation, climate of the site, the building usage and the technical installation. Design of the joints is also one of the major criteria as it controls the durability of the timber structures (Racher 1995). Racher (1995) also mentioned that the stability of the timber design is controlled by the design of the joints. Each of the joints should be able to transfer load from member to member without the connector material itself failing and without damaging the timber members in joints (Stalnaker and Harris 1997).

One of the most common types of joint in structural timber building is the mortise and tenon joint. A mortise is a rectangular hole cut in timber, and a tenon is a projecting piece of timber that is slotted or shaped to fit the mortise, and mostly at the end of a beam member in a structural building. An example of a mortise and tenon connection is shown in Fig. 1.2.

R. Hassan et al., *Timber Connections*, SpringerBriefs in Applied Sciences and Technology, https://doi.org/10.1007/978-981-19-2697-6_1

**Fig. 1.1** Example of timber joints

**Fig. 1.2** Mortise and tenon

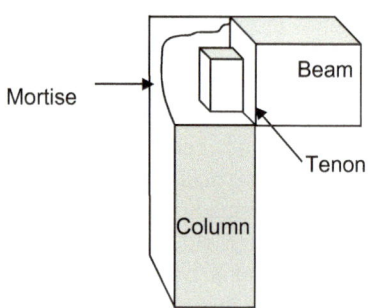

The mortise and tenon joint has been used in countless applications from furniture, to house construction, to ship building and has proven itself time and again (Schmidt and MacKay 1997). It is also proven that the mortise and tenon joints have survived for hundreds of years. In fact, at present, the only applied connection system for the new building construction system, namely the industrialised building system (IBS) for timber building is the mortise and tenon joint (Ismail 2009). Understanding the behaviour of mortise and tenon is also important as it also acts as a basis type of connections in all timber structural buildings.

Mortise and tenon with the aid of fasteners function to hold the shear, bending and tensile movement in the structure. Though it has been used for centuries, the guidelines to design the mortise and tenon is still not available in most main structural timber design standard. Due to the limited existing standard in designing mortise and tenon, timber connection designers normally utilise the common practice in the existing building structures and previous experience to predict and anticipate the capacity of the mortise and tenon joint. Thus far, the European Yield Model (EYM) is the closest reference used to predict the load-carrying capacity in the mortise and tenon joint design.

The prediction of load design capacity in the structural connections is very important for timber design engineers in order to produce a safe timber connection design. Estimation of the correct ability of each joint is a crucial input to the planning and design of the whole building structure. In the design contact, the understanding of capacity analysis can assist the timber joint designers to justify the best implementation in terms of strength, size and cost.

The common fastener for mortise and tenon is made of steel while the traditional type is made of wood dowel (Fig. 1.3). These dowels have their advantages and disadvantages. Though the steel dowel is well known for its stiffness, corrosiveness is still one of its major problems. Wood dowel has been used for centuries and proved to be strong and reliable in its own way and comes with its own weaknesses. However, wood dowel in mortise and tenon studies are so limited. Focus is in developing the structural timber in design, fundamental and in combination with the new material and technology was not very common.

Nowadays, glass fibre reinforced polymer (GFRP) is widely known for its structural uses. Its advantages are such as very stiff connections, particularly when the dowels are loaded in the axial direction, having a high local force transfer, good fibre properties, low cost material and production.

It is also improved aesthetics as the connections are completely hidden in the timber structure and light weight connectors in comparison with steel. GFRP has been recently explored for engineering applications such as in timber beams and bridge applications. Consequently, GFRP dowel applicability is evaluated and verified as a substitute to the steel dowel fastener for structural timber connections (Fig. 1.4).

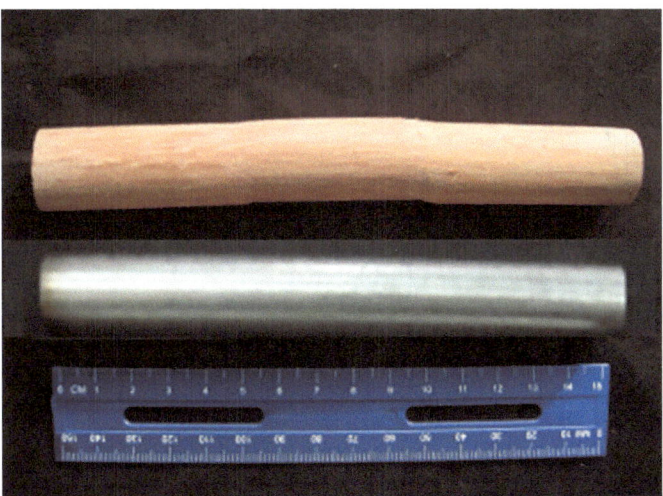

**Fig. 1.3**  Wood and steel dowels

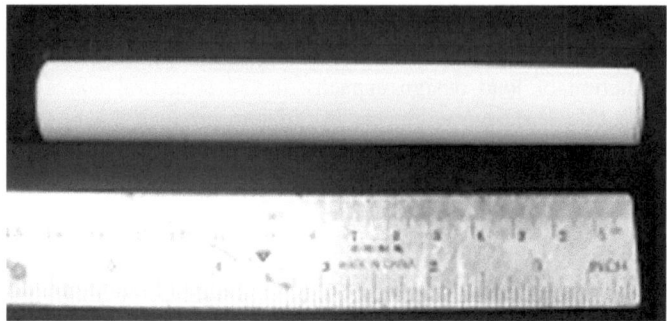

**Fig. 1.4** GFRP dowel

## 1.2   Basic Design of Mortise and Tenon

Mortise and tenon are well known as one of the traditional joints of timber construction and have been employed successfully over the period of time. However, the present design code of practice for timber—National Design Specifications 2005 (NDS 2005), Eurocode 5 (EC 5 2008) and Malaysian Standard (MS 544 2017)—does not include the design for mortise and tenon.

Among pioneered published guidelines in timber frame structures which include mortise and tenon is a standard produced by a group of structural timber researchers under the name of Timber Frame Engineering Council (TFEC). Located in Becket United States, under the work by Timber Framers Guild and the Timber Frame Business Council has produced a standard for Design of Timber Frame Structures and Commentary, namely TFEC 1-2007 and TFEC 1-2010, which incorporates a brief guideline on mortise and tenon connections loaded in shear and in tensile (TFEC 1-2007 and TFEC 1-2010).

This standard is developed as an intention to be as a supplement to provisions of the ANSI/AF&PA NDS 2005. However, as up to this date TFEC 1-2007 and TFEC 1-2010 standard has not been formally revised to be as a part of the NDS 2005. Therefore, the NDS 2005, EC 5, 2008 and MS 544, 2017 remain the primary governing design document for structural design of timber buildings. Nonetheless, few recommendations on timber design for mortise and tenon preparations and considerations from TFEC 1-2007 and TFEC 1-2010 were taken as references.

The European Yield Model (EYM) is a basis theoretical yield limit equation to determine the load-carrying capacity of timber joints applied in both well-known design codes, NDS 2005 and EC 5, 2008. The differences between these two codes are in their mathematical units and also on the experimental works used to develop the EYM. The EYM in the NDS 2005 code of practice as a basic theoretical development reference based on few reasons are discussed further in the next subsection.

In Malaysia, the mortise and tenon joint are constructed by the carpenters based on experience as the design code for structural connection of mortise and tenon is very limited. However, the design of timber joint using steel connectors is given

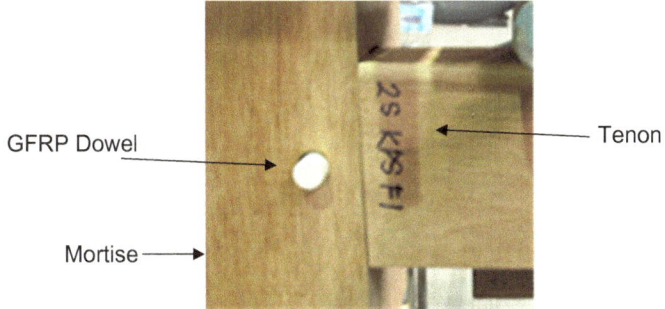

GFRP Dowel

Tenon

Mortise

**Fig. 1.5**  GFRP dowel functions to strengthen mortise and tenon joint

in MS 544: Part 5: 2018. The load-carrying behaviour of timber mortise and tenon joints made with Malaysian tropical timber with regard to rigidity, load capacity and ductility is not well-documented.

As for now, only scientist Ahmad et al. (1992) have reported on the performance of the mortise and tenon joint with wood dowels using Malaysian timber, namely Nyatoh (a species of *Sapotaceae*), Ramin (*Gonystylus spp.*) and Rubberwood (*Hevea brasiliensis*) for furniture manufacturing. The joint strength of mortise and tenon connection also depends on the width of the tenon, the width of the timber, the timber species, the size and the type of dowels. Since the existing types of dowels are made of steel, therefore the problems associated with fire resistance, corrosion and brittle splitting failures are all introduced through the use of steel fasteners and plates (Thomson 2009).

Hence, there is a strong incentive to evaluate the behaviour and strength properties of timber joint using different dowel materials which could possibly eliminate this problem. Particular emphasis is on the introduction of the GFRP into the design which acted as a new composite material as part of the fasteners strengthening system. Figure 1.5 shows the GFRP dowel is used to strengthen mortise and tenon joint.

The use of a composite material, GFRP, as substitute for steel has been recently explored for engineering applications for repair and strengthening of timber beams and bridges. Consequently, the GFRP dowel applicability as a substitute to the steel and wood dowel fasteners for timber connections, especially into the structural mortise and tenon type of joint is evaluated. However, the design using GFRP dowel is not available in NDS 2005; MS 544, 2017 nor is EC 5, 2008, and therefore, the suitability of the EYM was first observed in the use of GFRP dowel design capacity.

The observations on performance, capacity and behaviour of double shear, timber-to-timber joints dowelled with GFRP were compared with the performance of double shear, timber-to-timber joints dowelled with steel for validation purposes. Comparisons were also made to double shear, timber-to-timber joint dowelled with wood for further comparison and analysis. Few approaches to modify EYM for wood dowel with overseas timber are available for references, yet very limited research exhibits founding using tropical timber.

Many issues in EYM still need careful consideration. In order to use the EYM equation with tropical timber and to suit the mortise and tenon design prediction, the equations still need empirical adjustments and simplifications that best fit the local conditions practice. The EYM applicability to a large diameter dowel or other material-type fasteners made of material other than wood and steel also require further empirical observations. The EYM equation is then further explored to predict load-carrying capacity of actual structural mortise and tenon joints to meet the needs of timber design.

## 1.3  Conclusions

The purpose of this book is to share and discuss on details about the preliminary process and recorded history to the performance of GFRP dowelled mortise and tenon timber joints. It is focusing on the reliability of the EYM, NDS 2005 in estimating the load-carrying capacity of Kempas and Kapur species. The sharing knowledge in detail are related to dowel bending yield strength and dowel-bearing strength of wood properties; load-carrying capacity of double shear joint fastened with steel, GFRP and wood dowel and double shear joint fastened using steel and the related factor of safety in mortise and tenon design.

## References

S. Ahmad, A. Haji Amin, R. Ali, H. Md Tahir, Widthdrawal and bending strengths of dowels from three Malaysian timbers. J. Trop. For. Sci. **6**(1), 74–80 (1992)

Eurocode 5. Design of Timber Structures (BS EN 1995-1-1-2004 +A1:2008)

R. Ismail, Vendor Development Programme in Promoting the Industrialised Building System (IBS)-Implementation in the Construction Sector. Seminar on Opportunities and Prospects for Timber and Timber Products in the Domestic Market. PGRM Tower 2, Cheras. Kuala Lumpur, 28th May 2009.

MS 544: Part 2: 2017, *Malaysian Standard. Code of Practice for Structural use of Timber* (Department of Standards Malaysia. SIRIM, 2017)

MS 544: Part 5: 2018, *Malaysian Standard. Code of Practice for Structural Use of Timber—Part 5: Timber Joints*, 1st rev. Department of Standards Malaysia. SIRIM, 2018

J.L. Natterer, J.L. Sanders, Conceptual design. Timber Eng. **STEP 2**, E2 (1995)

NDS, *National Design Specification for Wood Construction* (American Forest and Paper Association (AFPA) Washington D.C., 2005)

P. Racher, Mechanical Timber Joints-General. Structural Timber Education Programme, Part 1. Lecture A3 (1995)

R.J. Schmidt, R.B. MacKay, *Timber Frame Tension Joinery*. Report for Timber Framers Guild. Becket, MA (1997)

J.S. Stalnaker, E.C. Harris, *Book of 'Structural Design in Wood* (Chapman and Hall, USA, 1997)

L. Stehn, H. Johansson, Ductility aspects in nailed glue laminated timber connection design. J. Struct. Eng. **128**(3), 382–389 (2002)

TFEC 1-2007, *Standard for Design of Timber Frame Structures and Commentary (Standard)* (Timber Frame Business Council and Timber Frame Engineering Council, (TFEC-TAC), Becket, 2007)

TFEC 1-2010, *Standard for Design of Timber Frame Structures and Commentary (Standard)* (Timber Frame Business Council and Timber Frame Engineering Council, (TFEC-TAC), Becket, 2010)

A. Thomson, R. Harris, P. Walker, M. Ansell, W.S. Chang, Contemporary metal free dowel connectors for timber structures, in *Proceedings of the 11th International Conference on Non-conventional Materials and Technologies (NOCMAT 2009)*, 6–9 Sept 2009. Bath, UK

# Chapter 2
# Mechanical Fasteners

## 2.1 Introduction

Traditionally, as a structural material, timber transferred the forces from one member to another by the construction of carpentry joints such as lap joints, cogging joints, tenon joints and framed joints (McKenzie and Zhang 2007) as indicated in Fig. 2.1. According to academicians McKenzie and Zhang (2007), the physical contact and the friction between timber members were relied upon to transfer the forces between them.

Timber connections are developed based on the type of mechanical connections and also the type of the connections. The mechanical joints are the types of fasteners used to strengthen these joints. The following section explains the most important behaviour in understanding the mechanical joints that is the ductility. Ductility behaviour and its evaluations are discussed in the preceding sections, followed by the introduction to the mechanical timber joint and the type of joints, which mainly focus on the mortise and tenon, respectively.

## 2.2 Ductility

Ductility is one of the variables used to determine behaviour in materials including nail joint capacity. It is also very important for structures in seismic regions (Blab and Schadle 2011). Any materials can be classified as ductile or brittle depending on its stress–strain characteristics. The ductility is when the material is subjected to large strains before it ruptures. Ductile materials are often chosen compared to brittle materials due to their capability of absorbing energy and will usually depict large displacement before failing (Hibbeler 2003). A ductile failure maintains a slightly higher strength capacity of the joints and must be preferred compared to a brittle failure which reduces the strength capacity of the joint (Stehn and Johansson 2002). Brittle behaviour is in contrast with ductile failure. A recent study for ductility for

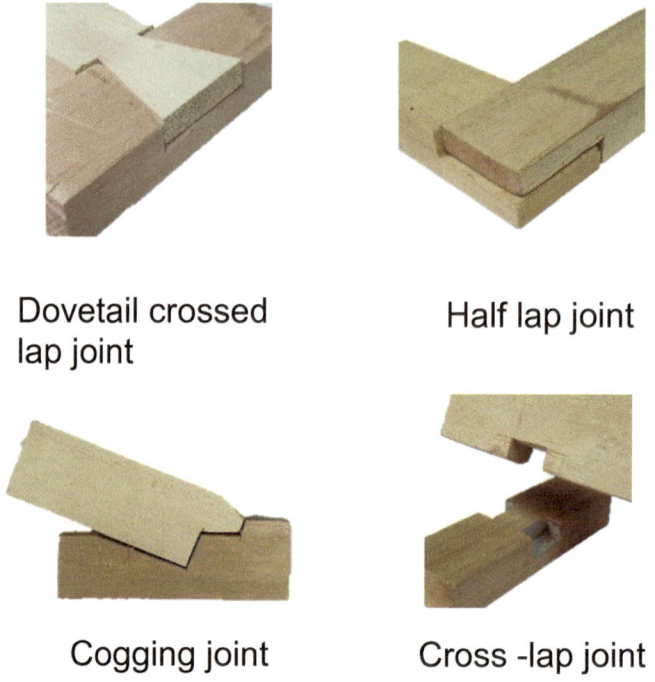

Dovetail crossed
lap joint

Half lap joint

Cogging joint

Cross -lap joint

**Fig. 2.1** Common type of carpentry joints

timber joints under cyclic loads has been investigated by Piazza et al. (2011). The ductility aspect of reinforced and non-reinforced timber joints using self-tapping screw for cross-laminated timber has also been studied by Blab and Schadle (2011).

## 2.3  Mechanical Timber Connections

Various modern methods are available in connecting the timber joints. Some are multipurpose fasteners, and some are specially produced for timber engineering Ozelton and Baird (2006). Many factors may influence the use of the particular type of the fastener, such as the method of assembly, connection details, purpose of connection, loading, permissible stress and aesthetics (McKenzie and Zhang 2007). These types of fasteners are described briefly as below:

(a)  Nails

Nails are suitable for lightly loaded connections and when they are loaded in shear. Normally, they are used in circular form as they are cut from wire coil (Ozelton and Baird 2006). There are many different types of nails manufactured from a wide variety of material, in many different shapes and sizes. The different types can be

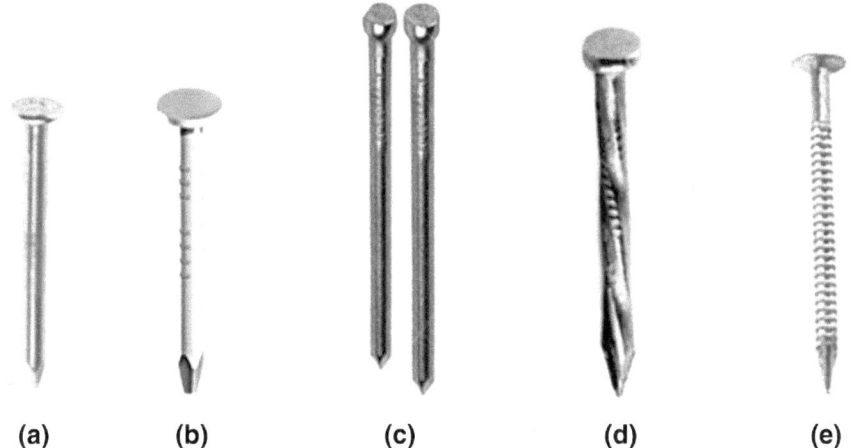

**(a)** **(b)** **(c)** **(d)** **(e)**

**Fig. 2.2** Example of nails **a** round plain nails, **b** clout nails, **c** lost head nails, **d** square twisted and **e** annular ringed shank nails

distinguished by characteristics such as nail head, shank, nail point, material type and the surface condition, for example, threaded or smooth (Harvey 2003). Example of nails is indicated in Fig. 2.2.

It can be treated against corrosion and increasingly being made of stainless steel for nail manufacture (Ozelton and Baird 2006). Smith et al. (1988) commented that in a single shear connection, the fastener is subjected to lateral loading. Under lateral loading, nail joints normally fail in a ductile manner, where a plastic hinge forms in the nails at the interface between the jointed members. This observation has been supported by Ozelton and Baird (2006), by stating that nails will slightly indent the timber when loaded in shear. Nailed connections are usually used for relatively small loads. Other types of fasteners such as bolts are normally used for larger loads (Harvey 2003). Nails joint can also be subjected to withdrawal loads such as wind uplift (Winistorfer and Soltis 1994).

(b) Staples

Normally used for lightly loaded connections between sheet material and solid timber. Its function is to give close contact during curing of a glued joint. Staples have a high stress concentration due to the forming of the staples at the shoulders, and therefore, it is not recommended to be used in an environment where corrosion of steel can occur (Ozelton and Baird 2006).

(c) Screws

Wood screws are usually used in steel-to-timber and panel-to-timber joints; however, it is also suitable for timber-to-timber joints. Screws tend to perform well under withdrawal loads, due to their treads (Harvey 2003). Conventional wood screws manufactured to BS EN 1210 require preboring and are therefore very much slower to insert compared to nails. Few types of with a modified thread pattern may be known as self-taping and do not require preboring. It can be found in various finishes or materials to cover different environmental conditions (Ozelton and Baird 2006).

(d)  Bolts and Nuts

Mild steel, either black or galvanised, is the ordinary bolts and dowels that are usually used in timber engineering. If a bolt is galvanised, the designer should ensure that the nut will still fit. If necessary this nut shall be 'tap' out or 'run down' the thread before galvanising. In corrosive environments, stainless steel bolts may also be used (Ozelton and Baird 2006). Bolts are normally used to strengthen the double or multiple joints. Soltis (1999) stated that they are few factors which influence the strength of a bolted connection. These include bolt and timber properties, member thickness, spacing, edge and end distances and bolts holes.

(e)  Dowels

Dowels are slender cylindrical rods, sometimes known as drift pin. They are commonly made of steel; however, non-metallic rods made of carbon fibre reinforced polymer (CFRP) or glass fibre reinforced polymer (GFRP) are widely available and increasingly being used. Dowels made of wood are also available and commonly known as pegs. Wood dowels are widely used to connect traditional timber joints such as mortise and tenon for structural or furniture joint purposes. Normally dowels are tightly fitted to the predrilled holes and are formed in a smooth surface. Harvey (2003) stated that in some cases, bolts are used with dowels, where the bolts are designed to take the tensile load and the dowels designed to take the shear load.

(f)  Shear Plates and Split Rings

Shear plates and split rings types of connectors need an anti-corrosion treatment. According to Ozelton and Baird (2006), shear plates are made of pressed steel or marble cast iron (Fig. 2.3). Joints using the shear plates will need to be cut out by a special cutter to take each connector, but no tool is required to draw the timber together. A small amount of joint slip is likely to take place. Shear plate connectors are located in place for transit with two locating nails.

A special tool to grove out the timber is needed when the split ring is used. The shape of the split ring and a special drawing tool is also required to assemble the unit under pressure. A split ring connector can carry a relatively high lateral load (Ozelton and Baird 2006). Its load-carrying capacity is also higher than double-sided tooth plate connectors, and they can be used more easily in very dense timber (McKenzie and Zhang 2007). These connectors are available in straight form or bevelled sides as depicted in Fig. 2.4. Precision in grooving and boring for these types of connectors is important as the needs to be extremely close-fitting to transfer the load efficiently (Racher 1995).

(g)  Punched Metal Plates

Whale (1995) described the punched plate as a galvanised steel sheet which can be made to a size and shape according to the connections requirements. Teeth are

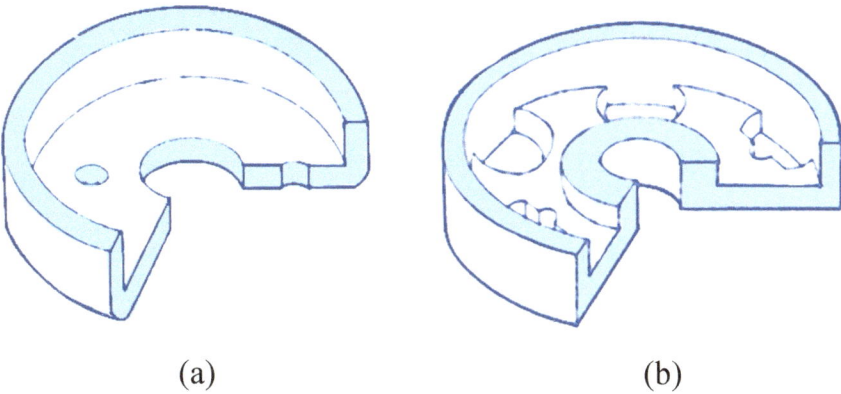

(a) (b)

**Fig. 2.3 a** Pressed steel shear plate and **b** malleable cast iron shear plate (Adopted from Ozelton and Baird 2006)

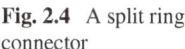

**Fig. 2.4** A split ring connector

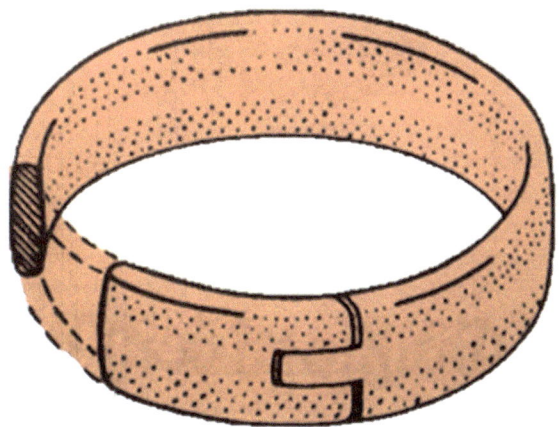

punched out of the plate and can be bent perpendicular to it. These teeth range from 0.01 to 0.10 in. in thickness and mostly used in prefabricated timber trusses. The important factors affecting the strength of joints dowelled with punched metal plates are the anchorage of the plate, and the tensile and shear strength properties of the steel.

(h) Toothed-Plate Connections

Toothed-plate connectors consist of circular plates of steel with toothed edges either a single-sided tooth plate or a double-sided tooth plate (Fig. 2.5). This type of connector is used in permanent joint and is embedded into the timber by the compressive action of the washers when the bolts are tightened.

**Fig. 2.5** Double-sided tooth
plate connector

## 2.4   Key Parameters Influencing Connections Capacity and Behaviour

Key parameters influencing connections shall be based on their capacity and their behaviour. As summarised by Harvey (2003), overall, the factors that affect the capacity of the mechanical connections are the material properties, moisture content, load duration, the dimension of the connector and timber members and number of fasteners for the multiple fasteners. Other related factors are the arrangement of the fasteners and the grain direction of the fastener and the members. Details factors affecting their capacity as explained by author Harvey (2003) are listed as follows:

- Material properties

The strength properties of the connectors are important, especially their yield strength and bending strength. The timber properties such as its embedment and bending strength are also the key considerations.

- Moisture content

Any fluctuation in moisture content may cause the timber to swell and shrink and loosen its connection. This may cause the loosening of dowel-type connections and lead to slight displacement as tolerance are taken up and embedment occurs.

- Load Duration

Timber experiences loss of strength and increasing deformation under a constant load over certain duration, known as creep. Creep tends to cause connections to become loose. The connecting material will also experience creep, but to a different extent depending on the material.

- Dimensions of the connector and timber members

Larger connector will take larger load, but the size of the timber member also needs to be considered.

- Arrangement of fasteners

Minimum spacing of the fasteners is as specified in NDS 2005. If the end distance, edge distance and spacing are too small, the timber will split.

- Direction of fastener force relative to grain

According to Soltis (1999), the joint deformation at the proportional limit is 30–50% more for loading perpendicular to the grain than for loading parallel to the grain. Perpendicular to the grain connections has also been identified as a priority critical to the advancement of joint design (Smith and Foliente 2002).

Further discussions on parameters that influence the behaviour of the connections were reported by Hein (2001). Hein (2001) specifically discussed on the fasteners yield strength and dowel-bearing strength as grouped by Harvey (2003) as part of the effects factor in the material properties.

## 2.5   Mortise and Tenon Connections

Mortise and tenon joints (Fig. 2.6) are very strong and most commonly found in structural timber buildings and fine furniture. Common types of mortise and tenon are the through mortise and tenon, open mortise and tenon and blind mortise and tenon (Figs. 2.7 and 2.8).

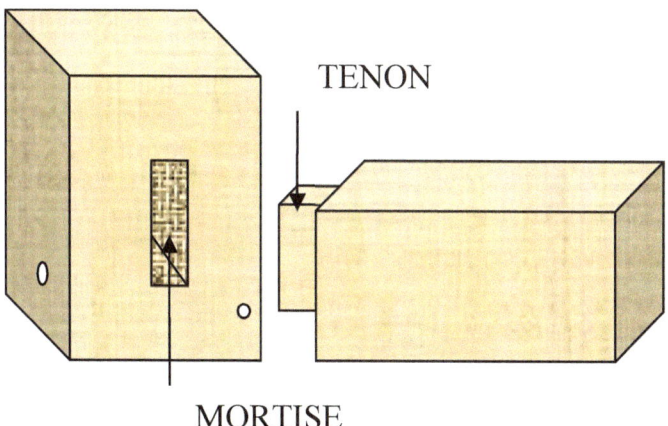

**Fig. 2.6** A mortise and tenon joint

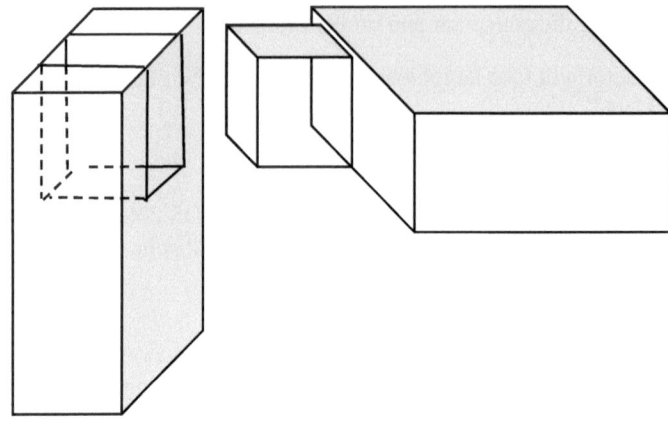

**Fig. 2.7**  Open mortise and tenon

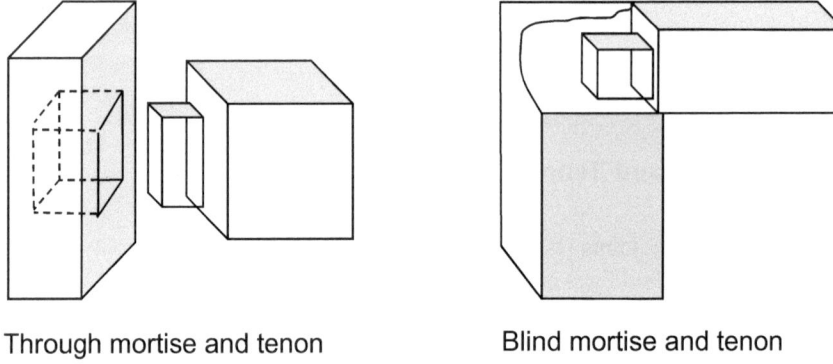

Through mortise and tenon                    Blind mortise and tenon

**Fig. 2.8**  Through and blind mortise and tenon

Eckelman (2006) and his group of researchers from Purdue University are actively involved in mortise and tenon joint studies. However, Eckelman is concerned more towards the use of mortise and tenon in furniture joints, and therefore, the sizes of his specimens were mostly smaller than the actual structural timber. Nevertheless, few recommendations and comments outcomes of his studies were referred to in some conditions such as on the effect of tenon shoulder on bending moment capacity for round mortise and tenon (Eckelman 2006).

Other researches are such as bending moment capacity of rectangular mortise and tenon for furniture joints (Eckelman 2006; Erdil et al. 2005). However, Eckelman (2006) mostly constructed the mortise and tenon joint with adhesive, purposely for furniture and does not encounter structural mortise and design. Publications, especially in the mortise and tenon for structural timber connections, were found to be very limited. There have been studies on using timber frames prior to the publications of a dissertation at Stanford University by academician Brungraber (1985).

Brungraber (1985) was the first person whom highlighted some concerns on the performance of mortise and tenon in a timber structural framing system. He studied the mortise and tenon using a wood dowel as an individual joint, together with full-scale frame testing, finite element analysis of joint behaviour and using formula in computer that incorporated connection behaviour. Brungraber (1985) found that the dowels and mortises failed before the tenons and also concluded that increasing the dowel diameter is the most effective way to increase the mortise and tenon connection's strength and stiffness. His project has directed the traditional timber frame process in the modern years.

Brungraber (1985) has also predicted shear stress and bending stress distributions along the wood dowel. Localised stress concentration of wood dowel was expected. The bearing stress on wood dowel under loading in a cross-section of a mortise and tenon joint is illustrated in Fig. 2.9. The stresses have been categorised into three phases, i.e. the early, middle and late ultimate (Fig. 2.10). He found that the analytical shear stress was very similar to the predicted distributions. The wood dowel was found to be too shallow and subjected to a very high bearing stress.

In Germany, timber frame research has been reported by Kessel et al. (1988) based on the reconstruction of an eight-storey timber frame. Kessel and Augustin (1995) concluded that oak wood dowel has sufficient strength for use in modern timber construction. Sensitivity to the strength and stiffness of mortise and tenon connections was shown by the additional numerical and physical testing on the behaviour of full timber frame structures by Bulleit et al. (1999), Erikson (2003). Erickson (2003) has also stressed out that the stiffness of a timber frame dowelled with wood is highly dependent on the stiffness of the individual wood dowel connections, and in order

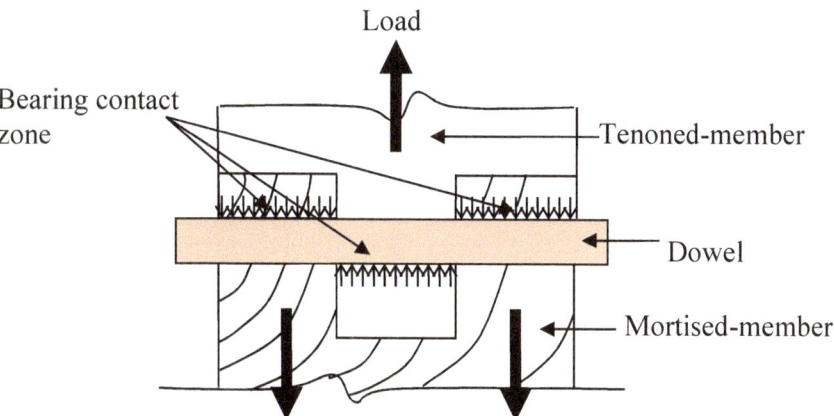

**Fig. 2.9** Bearing stress on wood dowel under loading in a cross-section of a mortise and tenon joint concept introduced by Brungraber (1985)

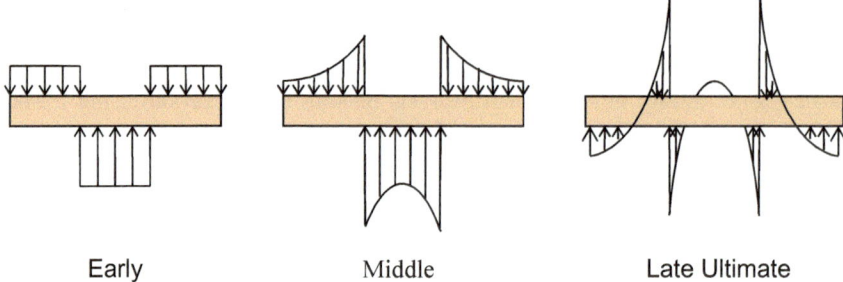

Early                        Middle                   Late Ultimate

**Fig. 2.10** Three phases of predicted bearing stress and distribution on the wood dowel (adopted from Brungraber 1985)

to accurately classify the displacement characteristics of a frame, the characteristics of the connections must be included in the assessment.

Upon this reason, therefore it is important to determine the performance and characteristics of each dowel since it has different stiffness according to its basic materials that is steel, GFRP and wood dowel in order to understand each joint characteristic. Since the design of traditional mortise and tenon fastened with hardwood is currently beyond the scope of building codes and the National Design Specification for Wood Construction (NDS 2005), Schmidt and his group of researchers from Wyoming University have studied on the applicability of the current EYM on the use of wood dowel (Schmidt and Mackay 1997; Schmidt and Daniels 1999; Miller 2004; Miller et al. 2010).

The research includes a study of the mechanical properties of the wood dowels used in the mortise and tenon connections under tensile and shear load. Properties of interest include the wood dowel's flexural yield strength, the dowel-bearing strength of a wood dowel as it loads the frame material, and the wood dowel's shear strength. The predominant loading component applied to this type of joint is said to be laterally loaded.

The results of their research show that the existing yield model equations (EYM) from the NDS 2005 are applicable to hardwood dowel used as dowel fasteners in mortise and tenon connections. However, they concluded that additional yield modes specific to these connections are needed. Schmidt and Mackay (1997), Schmidt and Daniels (1999), Miller (2004) proposed three (3) other additional yield modes to the existing EYM. The most current study was to establish a design method for the third additional yield mode for the wood dowel named as mode V. This research was done by Miller et al. (2010).

Investigation on mortise and tenon joints was investigated by other researchers with different parameters such as dowel diameter, end and edge distance of dowel, dowel orientation, bending and shear strength of dowels and bearing strength of wood dowels on base materials (Bulliet et al. 1999; Sandberg et al. 2000; Erikson 2003; Shanks and Walker 2005; Erdil et al. 2005; Walker et al. 2008).

Key authors in UK reporting on Green oak frame and box framing are Hewett and Harris (Hewett 1980; Harris 2001). They have reported on the historical primary connections of timber frames in UK. Yeoman has published on the repair of historic timber structures and consequently discussed details on traditional timber connections (Yeomans 2003). In the following years, Erikson (2003) at Wyoming University has examined the behaviour of traditional timber frame structures subjected to lateral load using unsheathed two-dimensional timber frames (Erikson 2003). Erikson (2003) has also developed viable stiffness-based methods and changing it to formula using the service-level behaviour of stand-alone timber frames subjected to lateral load.

Laboratory work of developing rational design guide lines for traditional joints in oak frame construction in UK has been studied by Shanks (2005). He has investigated structural performance of all wood dowel mortise and tenon joints and timber frame structures fabricated from oak to develop methods for analysing their strength and stiffness. Performance of mortise and tenon connections fastened with single wood dowel subject to shear, bending or tensile were investigated experimentally. As an outcome to their study, Shanks (2005), Shanks and Walker (2009) have proposed an energy method to predict the capacity of frame stiffness. This method is applied to predict the collapse loads of frames by equating the external work carried out in collapse to the energy dissipated as the frame collapse.

Listed below are the factors that are of concern for the mortise and tenon loaded in shear as published in TFEC 1-2007 and TFEC 1-2010.

(a) Load transfer by direct bearing—Transfer of shear load in a tenoned member to a mortise member shall be achieved by direct bearing of the tenon within the mortise or by direct bearing of tenoned member on a mortise housing
(b) Shear capacity—Shear load in a beam may be transferred to the mortised member through direct bearing across the width of the tenon for a beam that is not housed, or across the full width of the beam for a fully housed beam.
(c) Strength contribution of wood dowel—Wood dowels may be needed to carry some short-term load during assembly and erection of the frame, before the bearing surfaces come into full contact.

For the mortise and tenon loaded in tensile, more detail explanations were listed, which include the requirements for the yield limit equations, dowel-bearing strength, bending yield strength of the wood dowel, wood dowel diameter, edge distance, end distance and spacing, tenon size and quality and mortise placement. All necessary requirements as guided in this standard have been included in this study. Further information on the guidelines provided by the standard shall be referred directly to TFEC 1-2007 and TFEC 1-2010 Standard.

In Malaysia, though mortise and tenon are also known as one of the common traditional joint in Malaysian timber buildings, very rare publications concerning design and investigation of tropical timber were reported. In recent years, publications were limited to the work as exhibited in Malaysian Standard 544. Most timber houses in rural east and west Malaysia are still practising mortise and tenon as the main jointing systems. The famous historical timber buildings available in Malaysia are such as the old royal palace in Perak (Fig. 2.10) and a mosque in Kelantan (Fig. 2.11).

The Old Royal Palace is now a Historical Royal Museum in Kuala Kangsar, Perak. The museum building, originally built as a palace in 1926. It was made entirely from timber by a Malay builder on the royal command of His Royal Highness Sultan Iskandar Shah (1918–1938). Nowadays, only few original joints are exposed and

(a)

(b)

**Fig. 2.11**  **a** The Old Royal Palace in Kuala Kangsar, Perak, **b** an example of mortise and tenon jointing system creatively design

(a)

(b)

**Fig. 2.12** **a** The Masjid Kg. Laut, (A Mosque) in Kelantan, **b** an example of mortise and tenon jointing system applied at the mosque

most of the jointing system has been concealed under timber ceiling as part of beautifying the palace.

One of the most important living museums in Peninsular Malaysia made fully by timber is a mosque. Known as Masjid Kg Laut, which was built in 1930s, (Anon 2001) is located at Nilam Puri Kelantan and still actively used by the surrounding community as a religious centre. It was built using local timber and fully handcrafted. Mortise and tenon are the only type of structural joint applied for the mosque. The entire structure for this building was put together without hammering a single nail.

## References

Anon., Masjid Kampung Laut. Kelantan. Retrieved June 2010 from http://www.kelantan.gov.my (2001)

H.J. Blab, P. Schadle, Ductility aspect of reinforced and non-reinforced timber joints. J. Eng. Struct. (2011). https://doi.org/10.1016/j.engstruct.2011.02.001

R. Brungraber, *Traditional Timber Joinery: A Modern Approach*, Ph.D. Thesis, University of Stamford, 1985

W.M. Bulleit, L.B. Sandberg, M.W. Drewek, T.L. O'Bryant, Behaviour and modeling of wood-wood Dowelged timber frames. ASCE J. Struct. Eng. **125** (1), 3–9 (1999)

R.G. Erikson, *Behaviour of Traditional Timber Frame Structures Subjected to Lateral Loading*. Ph.D. Thesis, University of Wyoming, 2003

Y.Z. Erdil, A. Kasal, C.A. Eckelman, Bending moment capacity of rectangular mortise and Tenon furniture joints. For. Prod. J. **55**(12), 209–213 (2005)

C.A. Eckelman, E. Haviarova, H. Akcay, Exploratory study of the widthdrawal resistance of round mortise and Tenon joints with steel pipe cross pins. For. Prod. J. **56**(11/12), 55–61 (2006)

R. Harris, *Traditional Timber Framed Joinery*, 3rd edn. (Publications Ltd., Shire, 2001)

K. Harvey, *Improved Timber Connections Using Bonded-in GFRP Rods*. Ph.D. Thesis, University of Bath, United Kingdom, 2003

C. Heine, *Simulated Response of Degrading Hysteretic Joints with Slack Behaviour*. Published Ph.D. Thesis, Virginia Polytechnic Institute and State University, Blacksburg, Virginia, 2001

C.A. Hewett, *English Historic Carpentry* (Phillimore, London and Chichester, 1980)

R.C. Hibbeler, *Mechanics of Materials*, 5th edn. (Prentice Hall Publication, Upper Saddle River, New Jersey, 2003)

M. Kessel, R. Augustin, Load Behaviour of Connection with Oak Wood Dowels Timber Framing **38**, 6–9. Translation by R.J. Schmidt of: Untersuchungen uber das Tragverhalten von Verbindungen mit Eichenholzageln. Bauen mit Holz, 246–250 (April, 1995)

M.H. Kessel, M. Speich, F.J. Hinkes, The reconstruction of an eight-floor timber frame house at Hildesheim, in *International Timber Engineering Conference* (Forest Society, Madison Wisconsin, 1988), pp. 415–421

W.M.C. McKenzie, B. Zhang, *Book of Design of 'Structural Timber to Eurocode 5'*, 2nd edn. (Palgrave Macmillan, 2007)

J.F. Miller, *Capacity of Wood dowelled Mortise and Tenon Joints*, Master Science Thesis, Department of Civil and Architectural Engineering, University of Wyoming, Laramie Wyoming, 2004

J.F. Miller, R.J. Schmidt, W.M. Bulleit, A new yield model for wood dowel connections. J. Struct. Eng. (2010). https://doi.org/10.1061/(ASCE)ST.1943-541X.0000224

NDS, *National Design Specification for Wood Construction* (American Forest and Paper Association (AFPA), Washington D.C., 2005)

E.C. Ozelton, J.A. Baird, *Book of 'Timber Designers' Manual'*, 3rd edn. (Blackwell, 2006)

M. Piazza, A. Polastri, R. Tomasi, Ductility of timber joints under static and cyclic loads. Proc. ICE-Struct. Build. **164**(2), 79–90 (2011). ISSN: 0965-0911, E-ISSN: 1751-7702

P. Racher, Mechanical Timber Joints-General. Structural Timber Education Programme, Part 1. Lecture A3 (1995)

L.B. Sandberg, W.M. Bulleit, E.H. Reid, Strength and stiffness of oak wood dowels in traditional timber-frame joints. ASCE J. Struct. Eng. **126**(6), 21620 (2000)

R.J. Schmidt, R.B. MacKay, Timber Frame Tension Joinery. Report for Timber Framers Guild. Becket, MA, 1997

R.J. Schmidt, E.D. Daniels, Design Considerations for Mortise and Tenon Connections. Report for Timber Framers Guild. Becket, MA, 1999

J.D. Shanks, P. Walker, Experimental performance of mortise and Tenon connections in green oak. Struct. Eng. 40–45 (2005)

J.D. Shanks, *Developing Rational Design Guidelines for Traditional Joints in Oak Frame Construction*. Ph.D. Thesis. University of Bath, 2005

J. Shanks, P. Walker, Strength and the stiffness of all-wood doweled timber connections. J. Mater. Civ. Eng. ASCE. **1.21**(1), pp 10–18 (2009). ISSN 0899-1561/2009/

I. Smith, L.R.J. Whale, C. Anderson, B.O. Hilson, P.D. Rodd, Design properties of laterally loaded nailed or bolted joints. Can. J. Civ. Eng. **15**(4), 633–643 (1988)

I. Smith, G. Foliente, Load and resistance factor design of timber joints: international practice and future direction. J. Struct. Eng. ASCE **128**(1), 48–59 (2002)

L.A. Soltis, Fastenings Wood Handbook. Wood as an Engineering Material. Madison, WI, USDA Forest Service, Forest Product Laboratory. General Technical Report, FPL, GTR, pp. 7.1–7.28 (1999)

L. Stehn, H. Johansson, Ductility aspects in nailed glue laminated timber connection design. J. Struct. Eng. **128**(3), 382–389 (2002)

TFEC 1-2007, *Standard for Design of Timber Frame Structures and Commentary (Standard)* (Timber Frame Business Council and Timber Frame Engineering Council, (TFEC-TAC), Becket, 2007)

TFEC 1-2010, *Standard for Design of Timber Frame Structures and Commentary (Standard)* (Timber Frame Business Council and Timber Frame Engineering Council, (TFEC-TAC), Becket), 2010)

S.G. Winistorfer, L.A. Soltis, Lateral and withdrawal strength of nail connections for manufactured housing. J. Struct. **120**(12), 3577 (1994). https://doi.org/10.1061/(ASCE)0733-9445

L.R.J. Whale, Punched metal plate fastener joints. Timber Eng. **STEP 1**, C11/1–C11/8 (1995)

C.R. Walker, F.S. Fonseca, J.P. Judd, P.R. Thorley, Tension capacity of timber-frame mortise and Tenon connections, in *Proceeding of 10th World Conference on Timber Engineering (WCTE)*. Miyazaki, Japan, 2008

D. Yeomans, *The Repair of Historic Timber Structures*, 1st edn. (Thomas Telford, London, 2003)

# Chapter 3
# Structural Behaviour of Mortise and Tenon Joints

## 3.1 Introduction

The most common write up in this chapter are about mortise and tenon structural behaviour in bending moments, shear and tensile capacity compiled from the previous findings reported by related scientists and researchers in this field.

## 3.2 Bending Moments

A scientist, Brungraber (1985) classified the moment behaviour of mortise and tenon in two distinctive ways (Fig. 3.1). First, the wood dowels are taking the entire load when it is loaded in small capacity and rotation. Second, once the ends of the tenon contact the top and bottom of the mortise and begin to bear on each other and the wood dowels considered to have yielded enough thus the connections stiffen. In this second stage, the joint gradually stiffens as the bearing areas are relatively stuck and concentrated until the bearing points crush. According to the author, bending behaviour is a combination of tensile and compression behaviours.

Bending moments in mortise and tenon do have some resistance though it is commonly simplified as pin-jointed connections. The tenon member will rotate around the corner of the tenon shoulder once a bending moment is applied to a single wood dowel connection. Resistance to bending is provided by the lateral strength and stiffness of the wood dowel. It is acting at a lever-arm from the effective centre of rotation to the wood dowel centre line as shown in Fig. 3.2 (Shanks 2005).

Two scientists from Bath University, Shanks and Walker (2005) analysed the bending moment by supporting a mortise member on a strong wall and a load applied to the tenon. Prior to loading, the tenon member was propped and released to allow self-weight of the tenon member. The bending moment at which the dowel yields has been calculated from the ratio of the dowel to corner of tenon shoulder lever-arm and distance from the load point to the centre of rotation located at A (Fig. 3.3), creating

R. Hassan et al., *Timber Connections*, SpringerBriefs in Applied Sciences and Technology, https://doi.org/10.1007/978-981-19-2697-6_3

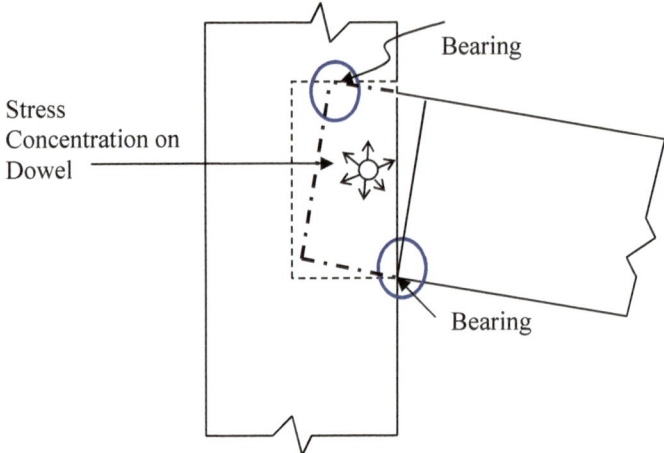

**Fig. 3.1** Bearing of tenon on top and bottom of mortise and stress concentration on dowel

**Fig. 3.2** Effective centre of rotation at corner of the tenon shoulder

an effectively solid hinge point. The tenon member will rotate around the corner of the tenon shoulder once a bending moment is applied to a single dowel connection. In their findings, the value of dowel yield moment for mortise and tenon is stated as 461 lbf.foot.

Moment rotation at A is determined by Eq. (3.1).

$$M_A = P_1 \times d_1 \qquad (3.1)$$

where $P_1$ is the value of force at a distance from the mortise face, $d_1$ is the perpendicular distance of load to the centre of rotation. Figures 3.4 and 3.5 show the schematic diagram and laboratory test set under bending load.

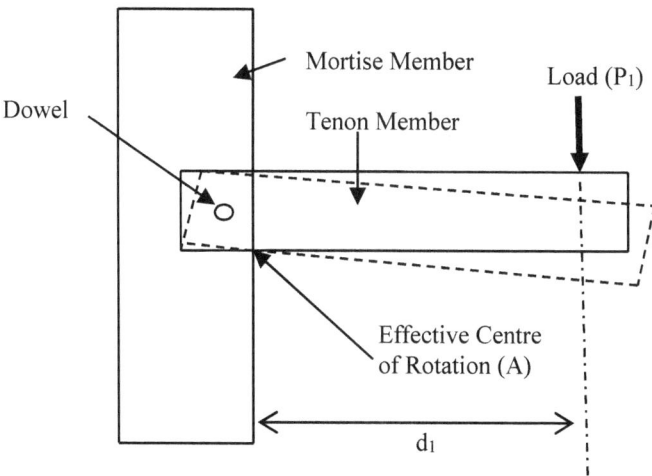

**Fig. 3.3** Schematic diagram of mortise and tenon joint subjected to bending load

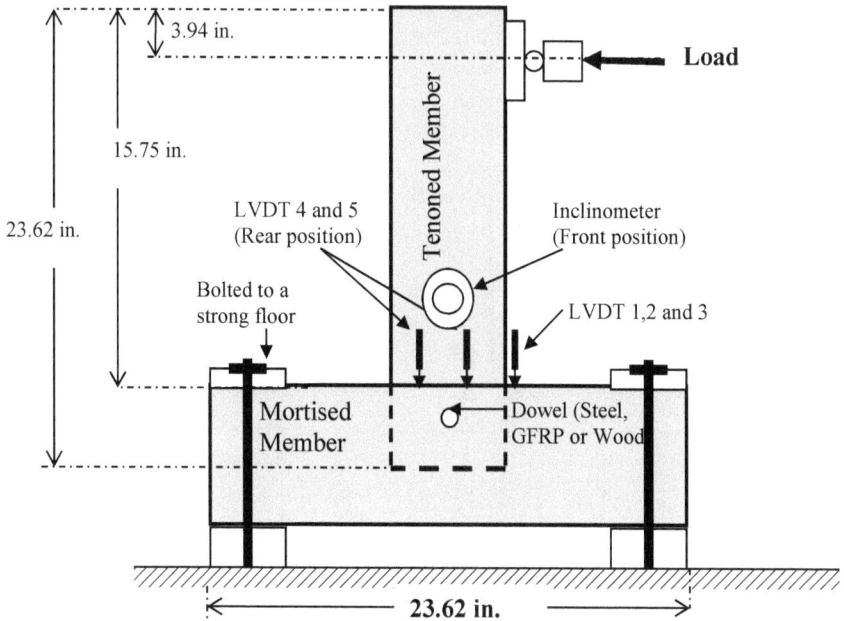

**Fig. 3.4** Schematic diagram of bending test set-up

**Fig. 3.5**  Experimental bending test set-up

Bending moment resistance of a rectangular tenon fits into a tight-fitting mortise and tenon using small specimens has been studied by Eckelman et al. (2006). In their case, the top and bottom of the tenon are supported by the tops and bottoms of the mortise; the sides of the tenon are bonded to the walls of the mortise. The strength of this joint is limited by the bending moment resistance of the tenon itself. The bending moment resistance has been determined by means of the expression in Eq. (3.2) and description in Fig. 3.6:

$$F_4 = \frac{td^2}{6} \times S_4,$$   (3.2)

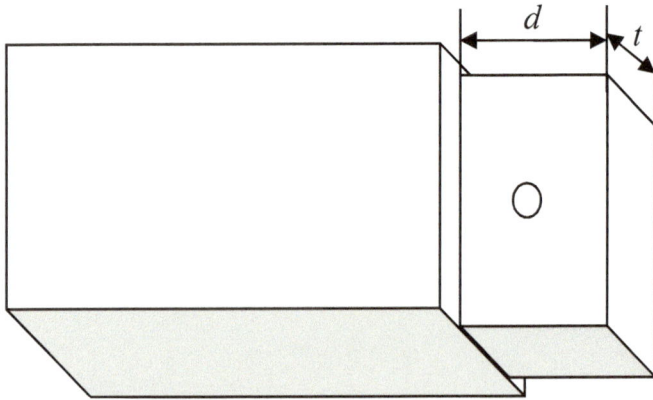

**Fig. 3.6**  Description of '$t$' and '$d$' of a tenon

where

$F_4$    the bending moment resistance of the tenon, kNm
$t$     thickness of the tenon, mm
$d$    the depth of the tenon, mm
$S_4$    the modulus of rupture of the material of which the tenon is constructed, N/mm$^2$.

The same paper also noted that this expression would apply primarily to those cases in which tenon depth equals member depth because in this condition, the value of bending moment resistance is at its maximum. For the bending moment capacities of a round tenon, as described by Wood Handbook, (1999) strength values can be estimated by means of the flexure using Eq. 3.3 (Wangaard, 1950):-

$$F_4 = \frac{k}{12} \times \frac{\pi D^3}{32} \times s_4, \qquad (3.3)$$

where

$F_4 =$   bending moment capacity (ft-lb)
$D =$   tenon diameter (in)
$s_4 =$   modulus of rupture (MOR) of the material (psi)
$k =$   a form factor for round beams, namely 1.18.

## 3.3  Shear Capacity

Shanks and Walker (2005) have analysed shear stress of green oak timber connection using full-sized of mortise and tenon with single timber wood dowel connection. Maximum applied shear loads for the connection ranged between 7686 lbf. and 15,961 lbf. Shear failure began as the tenon member cheeks 'rolled' from the tenon, which occurred between 2248 lbf. and 4496 lbf. The experimental work shows that the initial bearing area in shear between the mortise and tenon is very small, increasing only as the timber bearing crushes. Direct bearing is further reduced for both mortise and tenon which means that any shear load will initially be carried by the dowel in a similar mechanism to the tensile load, except with opposing grain directions. The mortised member in the opposite directions to the tenon will face more bearing stresses once the direct bearing is increased. Figures 3.7 and 3.8 show the schematic diagram and the laboratory set-up of the mortise and tenon test under shear load.

## 3.4  Tensile Capacity

Tensile capacity is also known as pull-out capacity of the mortise and tenon joint. The behaviour performance of this joint is also known as axial behaviour. According to

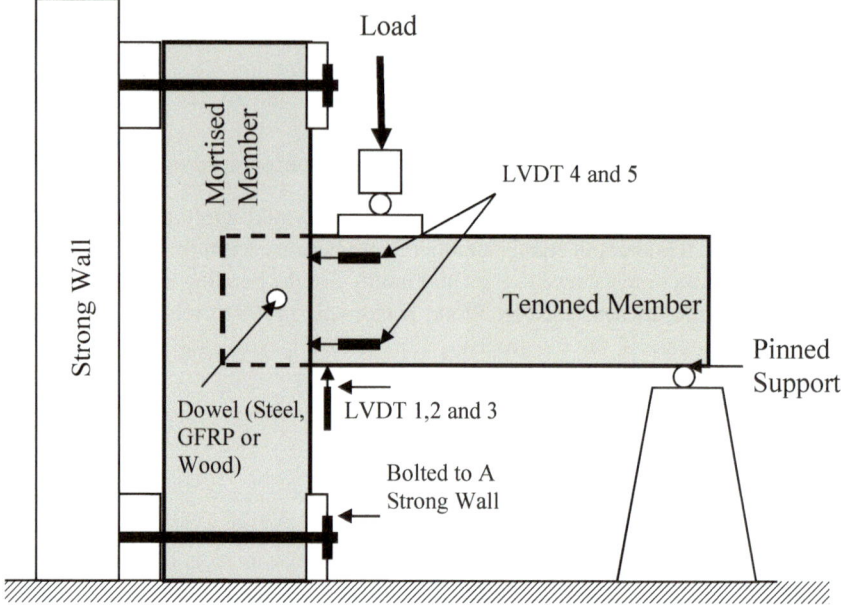

**Fig. 3.7**  Configuration of shear test set-up

**Fig. 3.8**  Experimental shear test set-up

Brungraber (1985), when the tenoned member is in tensile, the members will never bear solidly on the sides of the mortise. In fact, the wood dowel will simply continue to deform in shear and crush until the tenon pulls out of the mortise. Figures 3.9 and 3.10 show configuration and laboratory test set-up for a mortise and tenon under tensile load.

Most of the mortise and tenon joint were studied more in tensile (Brungraber 1985; Eckelman et al. 2004; Miller, 2004; Sandberg et al. 2000; Schmidt and Daniels 1999; Shanks 2005) rather than bending (Brungraber 1985; Eckelman 2003; Shanks 2005) and shear capacity (Brungraber, 1985; Miller, 2004; Shanks, 2005). Sandberg et al. 2000 have tested the mortise and tenon in tensile using the double shear or sandwich concept. This method was developed to represent the common mortise and tenon joints by arranging the middle pieces loaded in parallel to the grain and the side members loaded perpendicular to the grain.

Seventy-two (72) tests set-up were manufactured using eastern white pine or clear sugar maple with red oak dowels. Eastern white pine and clear sugar maple are in the range of 0.013 lbs/in.$^3$ and 0.025 lbs/in.$^3$, respectively, while European oak has

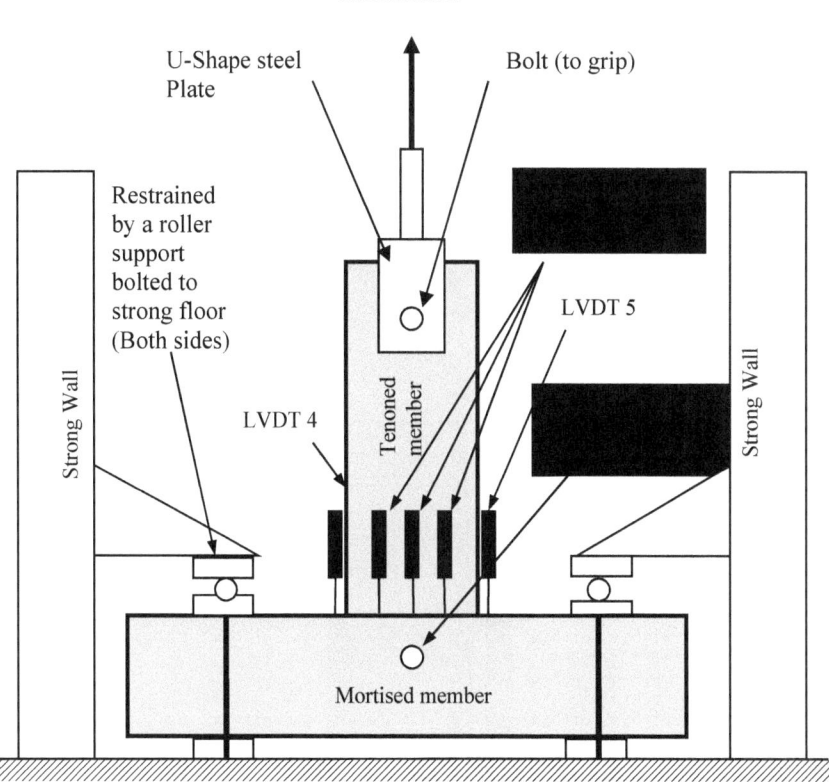

**Fig. 3.9** Configuration of experimental tensile test set-up

**Fig. 3.10** Laboratory tensile
test set-up

a density about 0.025 lbs/in.[3] (Sandberg et al. 2000). This range is common to US timber framing (Shanks 2005). The dowel diameter has been arranged in different principal of grain orientation, radially and tangentially loaded (Fig. 3.11).

Sandberg et al. (2000) found that the grain orientation has not affected the mortise and tenon joints. They also reported on the stiffness and strength of the mortise and tenon joints. For the analysis of joints stiffness, they found that the orientation of the perpendicular to the grain and the parallel to the grain direction of the middle and side plate only produced little influence. Sandberg et al. (2000) also found that the stiffness of the joint was affected by the member species but independent of grain orientation.

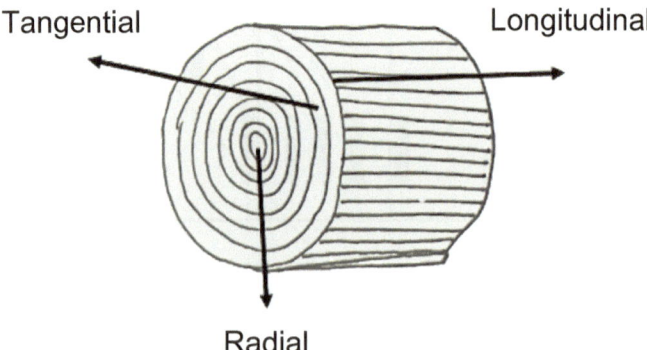

**Fig. 3.11** Principal of grain directions (tangential, radial and longitudinal)

For the analysis of joints strength, the experimental values resulted from their work have been compared with the modified EYM produced by Schmidt and Daniels (1999). The modified EYM consists of 6 equations proposed by Schmidt and Daniels (1999) and thoroughly discussed in Sect. 3.4 in this thesis. In all cases that Sandberg et al. (2000) reported, the EYM mode giving the minimum predicted strength corresponds to the observed failure mode. The EYM strength from the predictions was found on average 9.3% lower than the mean experimental values, and ranging from 15.8% lower to 1.9% higher depending on the species tested.

Sandberg et al. (2000) have also concluded that precaution of the gap between the mortise and tenon needs to be emphasised since this effect could be significant to the strength and stiffness of a mortise and tenon joint.

Shanks (2005) reported that the tensile tests for mortise and tenon for green oak have the average of 1686 lbf. as the ultimate load. The test was done by having the beam vertically restrained by holding down with roller supports to a strong floor at 47.24 in. centres. The tenon was loaded through a 0.98 in. steel dowel, through articulation to a load cell and hydraulic jack operated from a hand pump. All joints were found to have failed in tensile by withdrawal of the tenon following failure of the dowel.

Other than the capacity of the joint, the performances of the failure of the joints are normally observed. The failure mode of the joints dowelled with steel has been guided by the NDS (2005) which is also similar to EC 5, 2008. However, the failure modes for the joints dowelled with wood are not available in both standards. Nevertheless, as stated by Schmidt (2006), the tensile strength of a mortise and tenon joint connected with one or more wood dowels can be predicted using the yield formula approached in the NDS (2005).

## 3.5 Joint Design Using Malaysian Standard

The current standard for Malaysian in designing structural timbers is based on the Malaysian Standard 544 (MS 544: Part 2: 2017 and MS: Part 5: 2018). The Malaysian Standard committee is the Building and Civil Engineering Industry Standards Committee (ISC D) appointed by SIRIM Berhad. Development of this standard was carried out by the Construction Industry Development Board Malaysia (CIDB), which is the Standards-Writing Organization (SWO) appointed by SIRIM Berhad to develop standards for the construction industry. During the development of this standard, references were made to BS 5268: 1996. This standard is reviewed periodically. The newly revised version was made available in 2017 and 2018, respectively. The data from the various project and test results performed locally was also incorporated into the revised code (MS 544 2017).

MS 544 consists of several parts and sections, under the general title, 'Code of practice for structural use of timber'. These parts are:

Part 1:    General

Part 2:    Permissible stress design of solid timber
Part 3:    Permissible stress design of glued laminated timber
Part 4:    Timber panel products, i.e. structural and marine plywood, cement parti-
           cleboard and oriented strand board
Part 5:    Timber joints
Part 6:    Workmanship, inspection and maintenance
Part 7:    Testing
Part 8:    Design, fabrication and installation of prefabricated timber for roof trusses
Part 9:    Fire resistance of timber structures
Part 10:   Preservative treatment of structural timbers
Part 11:   Recommendation for the calculation basis for span tables for floor and
           ceiling joists, ceiling binders and rafters
Part 12:   Laminated veneer lumbers for structural application.

The most concern of this book is MS 544: Part 5: (2018), i.e. the timber joints section. For the purpose of joint design, the timber species has been classified into five joints groups: J1, J2, J3, J4 and J5. The joint classifications for the specific timbers are shown in Table 3.1. The selected species as an example of discussion in this book is Kempas and Kapur, classified into groups J2 and J3 accordingly.

Design of timber joints in Malaysia is based on the permissible stress load and currently limited to nails, wood screws, bolts, coach screws, split ring connectors and shear plate connectors. Tables of basic loads for each joint are applicable in the standard.

## 3.6  Strength Group and Grade Identification

As stipulated in MS 544: (2017): Part 2, Malaysian Timber has been classified into 7 strength groups (SG) of timber. The seven strength group is SG1, SG2, SG3, SG4, SG5, SG6 and SG7. The rank of these timbers is by having the strongest in structure and highest density in SG1 followed by decreasing strength and density by the increasing SG. Each strength group has its own numbers of species and is classified according to its stiffness and properties while the stiffness and properties of timber are according to its unique physical, mechanical or chemical properties.

For purpose of efficient utilisation either for structural or product uses, these species should be matched to end-use requirements through an understanding of their properties. This then requires identification of the species in timber form, independent of bark, foliage and other characteristics of tree. Even within each particular strength group, these species have different physical characteristics. Nevertheless, as describes by Handbook (1999) some species can be identified on the basis of readily visible characteristics such as colour, odour, density, presence of pitch or grain pattern (Anon, 1999). The physical properties may also vary within a species or be in different physical characteristics within one piece of timber. The existence of knots, shakes,

**Table 3.1** Joint classification of timbers for joint design as classified by MS 544: Part 5: (2018)

| Joint group | Strength Group | Timbers | | | |
|---|---|---|---|---|---|
| J1 | SG1 | Balau | Bitis | Chengal | Penaga |
| J2 | SG2/SG3 | Agoho<br>Dedaru<br>Kelat<br>Merbatu<br>Pauh Kijang<br>Ranggu | Bekak<br>Delek<br>Kembang semangkuk<br>Mertas<br>Penyau | Belian<br>***Kempas***<br>Kekatong<br>Mata ulat<br>Petaling | Balau<br>Keranji<br>Kulim<br>Perah<br>Surian batu |
| J3 | SG4 | Berangan<br>***Kapur***<br>Malabera<br>Meransi<br>Merpauh<br>Rengas | Dedali<br>Kasai<br>Meranti bakau<br>Nyalin<br>Resak | Derum Keruntum<br>Merawan<br>Perupok<br>Simpoh | Giam<br>Mempening<br>Merbau<br>Punah |
| J4 | SG5 | Alan Bunga<br>Brazil Nut<br>Kungkur<br>Meranti, Dark Red<br>Meranti, White<br>Ramin<br>Tembusu | Babai<br>Gerutu<br>Kelendang<br>Melunak<br>Nyatoh<br>Rubberwood<br>Teak | Balik angin bopeng<br>Kedondong<br>Keruing<br>Mempisang<br>Petai<br>Sepetir | Bintangor<br>Kayu kundur<br>Ketapang<br>Mengkulang<br>Penarahan<br>Sengkuang |
| J5 | SG6/SG7 | Ara<br>Damar<br>Minyak<br>Jenitri<br>Machang<br>Mersawa<br>Terap | Bayur<br>Geronggang<br>Kasah<br>Medang<br>Palajau<br>Terentang | Batai<br>Jelutong<br>Laran<br>Melantai/kawang<br>Pulai | Durian<br>Jongkong<br>Meranti, light red<br>Meranti, Sesendok |

checks and other imperfections will also affect the properties of timber (Pollack 1988).

Each strength group comprises of different timber stress grades. These stress grades are given in MS 544: Part 2: (2017), namely select structural, standard structural and common building grade with strength ratios of 80%, 63% and 50% of the basic stresses, respectively. For instance, a strength ratio of 63% means that the timber contains a defect which reduces its strength by 37% (Pun et al. 1997).

The species selected is based on its easiest in supply and among the common species available. The selected species are from SG2 and SG4. Since one of the main focuses was to validate the dowel-bearing strength and double shear strength of GFRP, steel and wood dowel using different species for the base material that holds the dowel therefore species with different strength groups were selected. Species of Kempas is representing the SG2, and Kapur represents the SG4, respectively. The characteristics of the selected species are referred to the MS 544: Part 2: (2017),

**Fig. 3.12** Texture of Kempas

Chik (1988a, 1988b); Gan et al. (1999) and Anon (2001) which are explained in the following subsections.

## 3.7 Kempas Species

Kempas is the standard Malaysian name for the timber of *Koompassia Malaccensis*. It is a hard and heavy timber with white and pale yellow sapwood. It can easily be distinguished from the brick red heartwood. The texture of the timber is coarse but even, with grains heavily interlocked with the presence of phloem. Generally, Kempas is difficult to work; however, it can be sanded to a good finish. Normal density for Kempas is 0.032 lbf./in.$^{3}$, categorised as medium hardwood in strength group 2 Malaysian Standard 544 (MS 544: Part 2: 2017). Kempas is not durable unless treated. Suitable for heavy construction, railway sleepers, transmission pots, beams, joists, bridges, wharves, fence posts, piling, parquet and strip flooring, panelling, heavy-duty furniture, heavy-duty pallets, boxes, crates and tool handles (Gan and Lim, 2004). Texture of Kempas species is shown in Fig. 3.12.

## 3.8 Kapur Species

Kapur is the standard Malaysian name for the timber of *Dryobalanops spp.* The trees of *Dryobalanops spp.* are generally tall and buttressed. It is relatively a freshly felled trees with camphor odour having scaly and sometimes flaky bole. The texture of the timber is coarse but even, with interlock grains. Generally, Kapur is easily

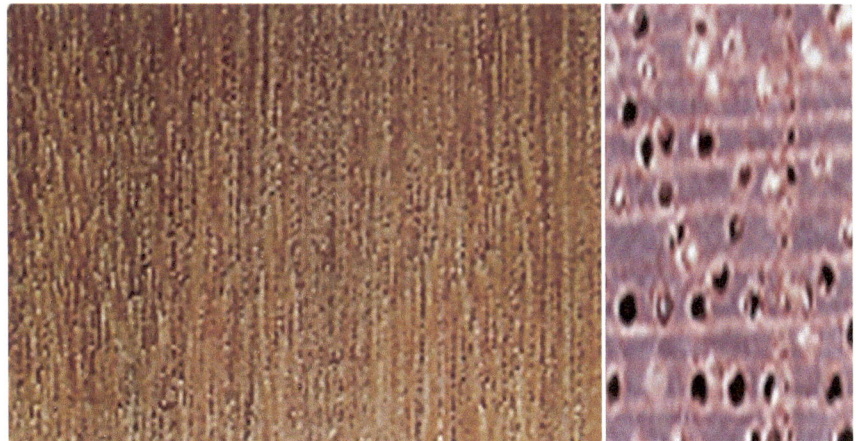

**Fig. 3.13** Texture of Kapur (Zainal et al. 2010)

to work with; however, slow drying and board before drying should be end-coated to avoid seasoning defects. Normal density for Kapur is 0.027 lbs/in.$^3$, categorised as a medium hardwood in strength group 4 Malaysian Standard 544 (MS 544: Part 2: 2001). It is relatively durable and suitable for heavy construction and medium flooring, beams, joists, rafters, furniture, doors and window frames, tool handles, pallets boxes and crates (Gan and Lim, 2004). Texture of Kapur species is shown in Fig. 3.13.

## 3.9 Characteristics of GFRP

GFRP also known as resin glass or fibre glass which arranged in unidirectional or two directions. Shen (2005) described GFRP as a good resistance to chemical when compared to other fibres. It also has low water absorption and therefore can be used for outdoor applications. It is made of a combination of polyester resin as a matrix and glass fibre as reinforcement, fillers and additives.

According to Shin (2003) glass fibres typically have a low modulus of elasticity when compared to carbon fibres or steel. However, steel has higher strain at failure which exhibits high ductility. This ductility gives advantage to steel over FRP (Dagher and Altimore, 2005). In terms of weight, the fibre content, resin and filler for GFRP are between 50 to 67%, 27 to 42% and 5 to 8%, respectively. The mechanical properties of glass fibre depend on fibre weight or volume content and fibre orientation (Jamaludin, 2002). Jamaludin (2002) has tested GFRP composites (type E-glass/polyester) and found that the tensile and compressive strength were 99.92 lbf./in$^2$. and 59.95 lbf./in$^2$., respectively, while the modulus of elasticity under tensile test was 5.94 lbf./in$^2$.

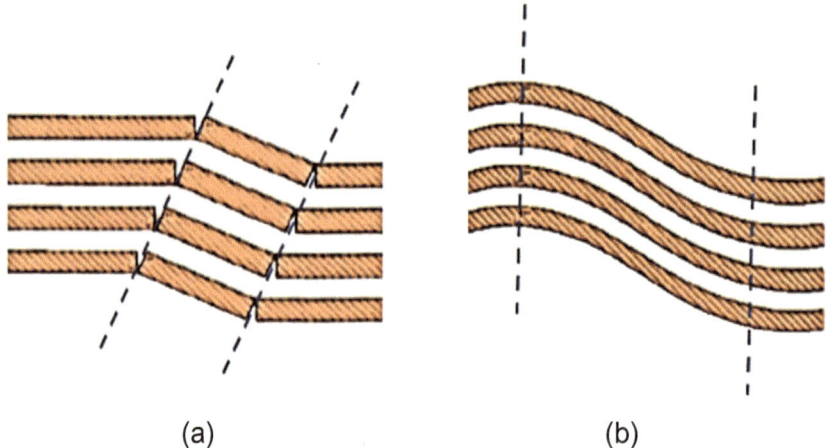

(a)                                                    (b)

**Fig. 3.14** **a** In-phase buckling leading to kink zone and **b** two planes of fracture from failure of carbon fibres adopted from Hull (1991)

In comparison with steel, GFRP is a composite with glass fibre bonded with resin while steel is more homogenous material. The nanofibres exist in GFRP dowel are also different and attributes to a coarser surface of GFRP dowel compare to the smoother surface of the steel dowel. Under bending, the failure of fibres causes tensile and compression stresses across the fibres due to in-phase buckling as shown in Fig. 3.14a. Brittle fibres, e.g. carbon fibres, form two planes of fracture as shown in Fig. 3.14b.

## 3.10  GFRP in Timber Applications

Plevris and Triantafillou (1992) with concerns on the tension zone of timber beams and beams columns using epoxy resins. The analysis is extended to establish a methodology for the optimum selection of the FRP reinforcement to optimise mechanical performance. Similar studies were done by Triantafillou (1997) and have studied the mechanical behaviour of timber members either reinforced or dowelled with fibre reinforced plastic (FRP) materials in the form of cloth (laminates or fabrics) externally bonded to the shear-critical zones. The author found that the result was quite satisfactory.

Fibre reinforced polymer (FRP) is composed from either carbon (CFRP) or glass fibre reinforced polymer (GFRP). The fibre reinforced polymer made of carbon (CFRP) is normally stronger and much expensive compared to the fibre reinforced polymer made of glass. FRP in structural applications has been explored in forms of dowel or in cloth-fibre. As an example, the fibreglass cloth was used to strengthen the single-bolted double shear timber connections by Soltis et al. (1997).

They found that the reinforced joints have increased their strength by 33% in comparison with the non-reinforced parallel to grain and more than twice the strength for perpendicular to grain stresses. Their findings also proved that the catastrophic failure associated with tensile-perpendicular to grain stresses was changed to a ductile mode of failure when the joints were bonded with the fibreglass cloth. In other related report, timber beams dowelled with GFRP dowels have been observed according to their development and applications by Gentile et al. (2002).

They studied the flexural behaviour of creosote-treated sawn Douglas fir timber beams connected with GFRP dowels. The results have shown that using the proposed experimental technique changed the failure mode from brittle tensile to ductile compression failure, and flexural strength increased by 18 to 46%. They also stated that due to the high strength of the GFRP reinforcement, the GFRP bars did not rupture in any of the specimens. Research findings indicate that the use of near-surface GFRP dowels overcomes the effect of local defects in the timber and enhances the bending strength of the members.

Other works using the cloth-fibre made of carbon fibre reinforced polymer (CFRP) are the flexural strength of timber structures by Schober and Rautenstrauch (2005) from Germany. From the findings, the arrangement of the reinforcement, the bond surface quality and the stiffness of the load transmitting materials were of decisive influence for the overall strength of the specimen.

Timber members dowelled with mechanically fastened with cloth-FRP has also been reported by Dempsey and Scott (2006). Their findings supported the Schober and Rautenstrauch (2005) by stating that the proposed strengthening technique induced a gradual failure of the composite members and increased ultimate moment, initial stiffness, and ductility over that found for the control specimens. Dempsey and Scott (2006) increased the fastener spacing and found that it decreased the member ultimate moment, initial stiffness and ductility ratio. The moisture content of the timber material greatly affected the ductility ratio of the timber members.

Research by Gentile et al. (2002) was supported by Ahmad (2010), who investigated the behaviour of Yellow Meranti beams strengthened with GFRP bars and CFRP plates bonded using Sikadur-30 as the bonding agent (Fig. 3.15). It was found that the stiffness, strength and ductility of the strengthened beams performed better than the un-strengthened beams.

Tests for GFRP dowelled in double shear joints manufactured using densified veneer timber plates were examined by Thomson et al. (2010). The GFRP dowels were of 0.47 in. diameter. They found that the GFRP dowels demonstrated favourable capacity when compared to stainless steel. Joints connected with GFRP dowel exhibit a very ductile behaviour at yield capacity. Due to this result, it was suggested that GFRP dowels are appropriate for further improvement (Thomson et al. 2010). A recent publication using CFRP for glued laminated timber (glulam) using glued joints has been reported by Juvandes and Barbosa (2011). They have proposed some recommendations for the design of timber structures reinforced with CFRP system.

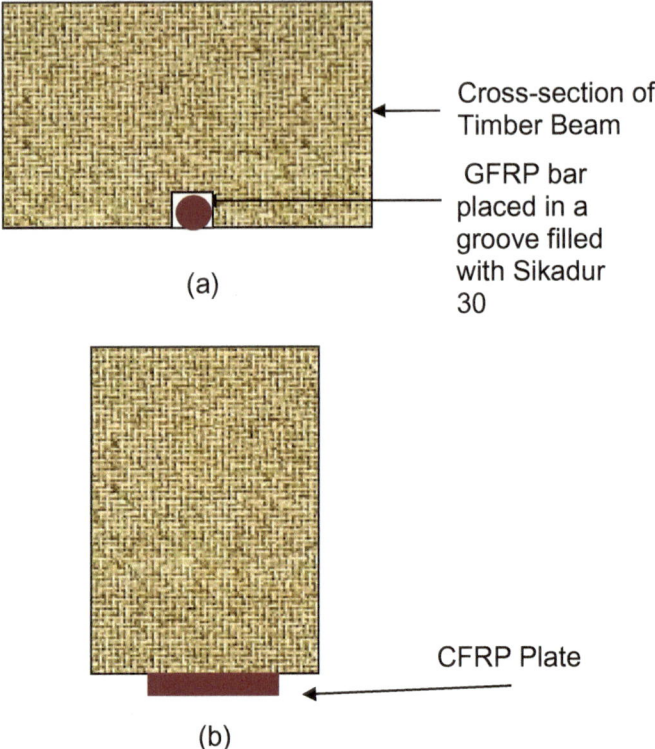

**Fig. 3.15  a** Applications of GFRP bar in timber beam and **b** application of CFRP plate in timber beam

## 3.11  Conclusions

Previous and current information about timber connections were covered in this chapter. It can be concluded that the development of timber connection is continuously under investigations since the standard for timber connection design is still limited. In order to understand the timber joints capacities and behaviour, the key parameters that influence the mechanical timber connection capacity and joints ductility behaviour were also discussed in this chapter.

Reviewed on the mortise and tenon joints were also covered as it is the main focus of this book. This type of joint was reported as the most common type of structural timber joint, nevertheless very limited sharing was found available. Hence, the information of mortise and tenon with various types of fasteners in different loading conditions is therefore important for timber structural applications. The structural capacity and behaviour of mortise and tenon joints were reported in this chapter based on the loading condition which includes bending, shear and tension condition.

Timber joint design covered in the Malaysian standard and information on the selected timber species in this book were also briefly explained. The characteristic of the GFRP and its current application in the timber engineering were also observed. From the related available information, it can be concluded that the use of GFRP in structural engineering is very promising and was reported compatible with other engineering construction materials.

# References

Y. Ahmad, *Bending Behaviour of Timber Beams Strengthened using Fibre Reinforced Polymer Bars and Plates*. Ph.D. Thesis. Faculty of Civil Engineering. (Universiti Teknologi Malaysia. Johor. Malaysia, 2010)

Anon, *Masjid Kampung Laut. Kelantan*. Retrieved June 2010 from http://www.kelantan.gov.my (2001)

R. Brungraber, *Traditional Timber Joinery: A Modern Approach*, Ph.D. Thesis, University of Stamford (1985)

E.A.R. Chik, Malaysian timbers—Kempas. Malaysia forest service. Timber leaflet No. 44, Forest Department of Peninsular Malaysia, Kuala Lumpur, Malaysia (1988a)

E.A.R. Chik, Malaysian timbers—Kapur. Malaysia forest service. Timber leaflet No. 46, Forest Department of Peninsular Malaysia, Kuala Lumpur, Malaysia (1988b)

H.J. Dagher, F.M. Altimore, Use of glass-fiber-reinforced polymer tendons for stress-laminating timber bridge decks. J. Bridg. Eng. **10**, 21–27 (2005)

D.D. Dempsey, D.W. Scott, Wood members strengthened with mechanically fastened FRP strips. J. Compos. Constr. **10**(5), 392–398 (2006)

C.A. Eckelman, E. Haviarova, H. Akcay, Exploratory study of the widthdrawal resistance of round mortise and tenon joints with steel pipe cross pins. For. Prod. J. **56**(11/12), 55–61 (2006)

K.S. Gan, K.T. Choo, S.C. Lim, Timber notes—medium hardwoods I–Kapur, Kasai, Kelat, Keledang & Kempas. *Timber Technology Bulletin* No. 11, Forest Research Institute Malaysia, Kepong (1999), p. 8

K.S. Gan, S.C. Lim, Common commercial timbers of peninsular Malaysia. Forest Product Technology Division. Forest Research Institute Malaysia (FRIM). Malaysia (2004)

C. Gentile, D. Svecova, S.H. Rizkalla, Timber beams strengthened with gfrp bars: development and applications. J. Compos. Constr. **6**, 1–71 (2002)

W. Handbook, *Wood as an Engineering Material* (United State Department of Agriculture, 1999)

D. Hull, *An Introduction to Composite Materials*. (Cambridge University Press, Cambridge, 1991). ISBN 9780521388559.

Y. Jamaludin, *Performance of Pultruded GFRP Composites under Tropical Climate*. (Doctor Philosophy, UTM, Skudai, 2002)

L.F.P. Juvandes, R.M.T. Barbosa, Bond analysis of timber structures strengthened with FRP system. Strain (2011). https://doi.org/10.1111/j.1475-1305.2011.00804.x

J.F. Miller, *Capacity of Wood dowelled Mortise and Tenon Joints*, Master Science Thesis, Department of Civil and Architectural Engineering, University of Wyoming, Laramie Wyoming (2004)

MS 544: Part 2, Malaysian Standard. Code of Practice for Structural use of Timber. Department of Standards Malaysia. SIRIM (2017)

MS 544: Part 5, Malaysian Standard. Code of practice for structural use of timber—Part 5: timber joints (First revision). Department of Standards Malaysia. SIRIM (2018)

NDS, National Design Specification for Wood Construction American Forest and Paper Association (AFPA) Washington D.C., (2005)

N. Plevris, T.C. Triantafillou, FRP-reinforced wood as structural material. J. Mater. Civ. Eng. **4**, 300–317 (1992)

H.W. Pollack, Material Science and Metallurgy, Wood and Paper (4th ed.). (New Jersey, 1988), pp. 486–492

C.Y. Pun, H.K. Seng, M.S. Midon, A.R.A. Malik, Timber Design Handbook. FRIM: Malayan Forest Records No. 42 (1997)

L.B. Sandberg, W.M. Bulleit, E.H. Reid, Strength and stiffness of oak wood dowels in traditional timber-frame joints. ASCE J. Struct. Eng. **126**(6), 21620 (2000)

R.J. Schmidt, E.D. Daniels (1999) Design Considerations for Mortise and Tenon Connections. Report for Timber Framers Guild. (Becket, MA)

R.J. Schmidt, Timber wood dowels consideration for mortise and tenon joint design. Wood Design Focus. **14**(3), 44–47 (2006)

K.U. Schober, K. Rautenstrauch, Experimental investigations of Fleural strengthening of timber structures with CFRP, in *Proceedings of the International Symposium on Bond Behaviour of FRP in Structures (BBFS 2005)*. (International Institute for FRP in Construction, 2005), pp. 457–464

J.D. Shanks, *Developing Rational Design Guidelines for Traditional Joints in Oak Frame Construction*. Ph.D. Thesis. University of Bath (2005)

J.D. Shanks, P. Walker, Experimental performance of mortise and tenon connections in green Oak. Struct. Eng. (6), 40–45 (Sept 2005)

C.H. Shen, *Prestasi Lenturan Rasuk Konkrit Bertetulang yang Diperkuatkan Dengan Plat Keluli dan Fabrik GFRP*. (Bachelor in Civil Engineering, UTM, Skudai, 2005)

F.T. Shin, *Kelakuan Lenturan Rasuk Komposit GFRP Pultruded*. (Bachelor in Civil Engineering, UTM, Skudai, 2003)

L.A. Soltis, R.J. Ross, D.F. Windorski, Effect of fibreglass reinforcement on the behaviour of bolted wood connections. J. Contemp. Wood Eng. **8**(3), 19–24 (1997)

A. Thomson, R. Harris, P. Walker, M. Ansell, *Development of Non-metallic Timber Connections for Contemporary Applications*. CD Proceeding of 11th World Conference on Timber Engineering (WCTE). 20–24 June 2010. Trentino. (Italy, 2010)

T.C. Triantafillou, Shear reinforcement of wood using FRP materials. J. Mater. Civ. Eng. **9**, 65–69 (1997)

F.F. Wangaard, *The Mechanical Properties of Wood* (Wiley, New York, 1950)

S. Zainal, H. Abdullah, A. Kong (2010) Book of 'On-site identification of some common timbers used in Malaysia'. Published by Malaysian Timber Council. ISBN 978-983-99314-9-5

# Chapter 4
# Load-Carrying Capacity of Timber Joints

## 4.1 Introduction

This chapter explains the mathematical and basic theoretical application as stipulated in the EYM, NDS (2005) (Fig. 4.1) and discussed on their limitations. Some references were also made to EC 5 (2008) for further understanding of the load-carrying capacity of the timber joints. The focus is on the timber-to-timber EYM equations and on the EYM modification for wood dowel application. The subsidiary literatures, that is the dowel-bearing strength of wood-based material as complimentary to further understand the EYM were also discussed. The previous studies on the influence parameters to the dowel-bearing strength are also covered. The discussion followed by the concept of factor of safety applied in this sharing ended by the conclusion of this chapter.

## 4.2 European Yield Model (EYM)

Load-carrying capacity of mechanical timber joints is predicted using European Yield Model (EYM) equation. It has been accepted globally as an engineering approach in joint design since it has been developed by Johansen (1949). Johansen (1949) who considered a number of possible failure modes relating to single shear, double shear, timber-to-timber, timber-to-panel and steel-to-timber connections. EYM afterwards has been expanded by Larsen (1973). Few assumptions were made by Larsen (1973) to analyse the dowel yield theory. It is assumed that the timber and dowels are elasto-plastic which allows the simplified static equilibrium expression (Fig. 4.2).

Other assumptions made are that the dowel is prismatic and bears on joint members uniformly across its diameter. For the double shear joints, it is considered as symmetrical. The dowels are also expected to be predrilled with holes that are slightly smaller than the dowels (Mackenzie and Zhang (2007). This expectation is different to Harvey (2003) as stated in Sect. 2.3 who's mentioned that normally dowels are

R. Hassan et al., *Timber Connections*, SpringerBriefs in Applied Sciences and Technology, https://doi.org/10.1007/978-981-19-2697-6_4

**Fig. 4.1**  National design specification book of reference (2005 edition)

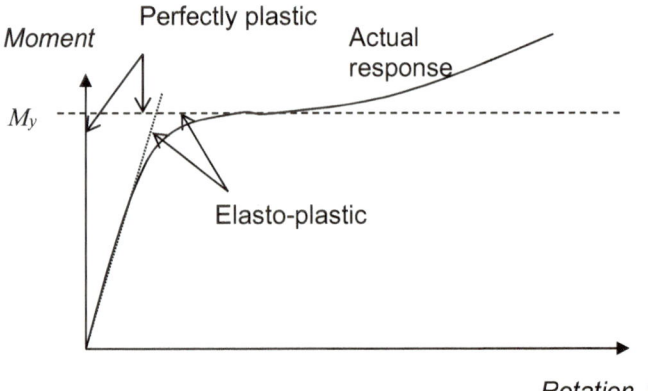

**Fig. 4.2**  Material behaviour assumed by Larsen (1973) and Shanks (2005)

tightly fitted to the predrilled holes. However, statement made by Harvey (2003) is for the common timber joints, while the statement made by Mackenzie and Zhang (2007) was specifically for the double shear type of joints.

EYM also does not account for friction between the joint members. The results of this work are reflected in the code in terms of equations from which the characteristics load-carrying capacity for nails, staples, bolts, dowels and screws per shear plane per fastener. The load-carrying capacity is determined for each possible mode of failure; the minimum value being used for design (McKenzie and Zhang 2003).

Later, the formula has been established and modified by Aune and Patton-Mallory (1986), Heine (2001), McLain and Thangjitham (1983), Miller (2004), Miller et al. (2010), Schmidt and Mackay (1997), Schmidt and Daniels (1999), Schmidt (2006), Soltis and Wilkinson (1987), Smart (2002) and Smith et al. (2005) to suit the common timber connection design (Fig. 4.3).

McLain and Thangjitham (1983) pointed out on the limitation of the EYM for developing design values. The several limitations are:

1. The displacement in a joint is not corporate in the EYM equations. However, the deformation in a connection could be an important issue as a limiting load.
2. Joints are only assumed to fail due to lateral load and not axial load. End distance and edge distance are also considered sufficient to prevent failure due to shear and splitting. McLain and Thangjitham (1983) supported the analytical methods as proposed by Wilkinson and Rowlands (1979) in finding the unknown exceptions to these observations, such as more appropriate spacing.

**Fig. 4.3** Common timber connection design

3. A modification of the equation in terms of theoretical to the practices is needed since the formula or the produced model assumed that the joint is well manufactured. The bolt is assumed to fit the holes in nearly perfect condition. It is thought that the modifications to the decision rules will compensate for usual design tolerances (NDS 2005).
4. Friction is ignored in the equations since it is unpredictable and will change with time and the environment experienced by the joint.
5. Rational compensation for safety is needed in order to apply the model predicted load in a working stress design.

Above all, the key affected parameters in designing using EYM are the diameter and yield strength of the fastener and the thickness of the joint members. The design is only for the single-dowel-type fasteners such as nail, screws, dowel, drift pin or bolt. Both single shear and double shear are using a similar theory of single-dowel yield mode. According to Heine (2001), any analytical analysis of multiple-dowel connections should start with the single fastener. Smith et al. (2005) reported that the strength of a joint with single fastener is applied by all countries as basis for assigning characteristic properties to determine the EYM load-carrying capacity. However, the yield points from the load–displacement curve, methods and mathematical units to derive the load-carrying capacity are not similar in some cases from the USA (NDS 2005) and Europe (EC 5 2008). The 5% diameter offset yield and the ultimate yield load is as depicted in Fig. 4.4.

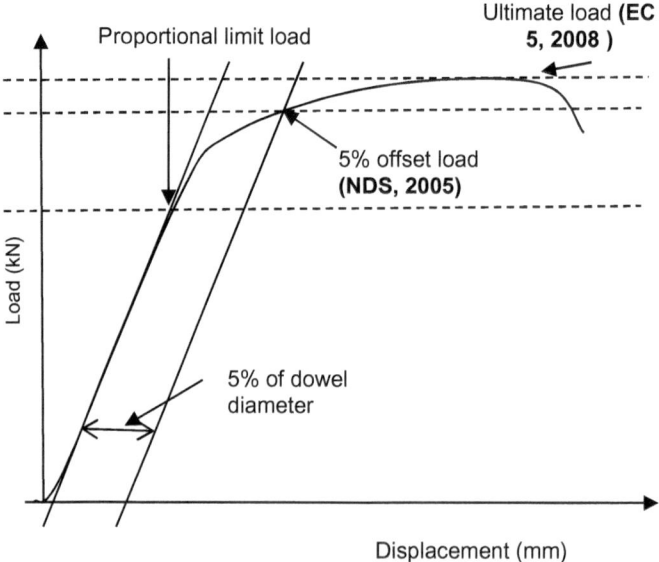

**Fig. 4.4** Illustration of the 5% diameter offset method and the ultimate load from load–displacement curve

In NDS (2005), all parameters of interest that include dowel-bearing strength (ASTM D5764), dowel bending yield moment (ASTM F-1575) and double shear joints in USA (ASTM D-5652-95 (reapproved 2007) and Mclain and Thangjitham 1993) are taken at 5% diameter offset yield, whereas for EC 5 (2008), the dowel-bearing strength (BS 383), dowel bending yield moment (BS EN 409) and the shear loaded joints are taken at ultimate or maximum load (Jumaat et al. 2008). Imperial units are being used in USA while matrices units are applied in EC 5 (2008).

However, the 5% diameter yield and the ultimate strengths of bolted timber joints can be estimated by the EYM theory when the appropriate dowel-bearing strength of wood and the bending yield moment of bolt are applied (Sawata and Yusumura 2002). The EYM is also the basis for the Canadian, Australian and New Zealand Code (Soltis and Wilkinson 1991).

The use of 5% offset in measuring the input properties for EYM was applied in many previous timber joint studies. Wilkinson (1993) found that the 5% offset yield load is generally less variable than either proportional limit or ultimate capacity. Schmidt and Daniels (1999) explained that while the use of 5% offset method is somewhat subjective, it is believed to be a more accurate and repeatable method for determining joint strength than the use of the proportional limit or ultimate strength. Their statement has been supported by Finkenbinder (2007) and Smith et al. (2005). Smith et al. (2005) found that the characteristics for one-dowel design capacities of joints are most reliably based on the yield load rather than the ultimate load.

Finkenbinder (2007) stated that between 5% offset and ultimate capacity, a majority of validation studies have been focused at the 5% offset yield level, with less consideration behaviour given to behaviour at ultimate capacity (Finkenbinder 2007). Finkenbinder (2007) also commented that further research of connection behaviour at ultimate capacity is in need to validate the application of the general dowel equation at ultimate capacity. Currently, only 5% offset yield is of concern for all ASTM Standard, while British Standard (BS) tends to use ultimate load rather than the 5% offset to determine the yield capacity for joint strength.

## 4.3 National Design Standard (NDS 2005)

As stipulated in NDS (2005), the different failure modes of EYM for the single and double shear timber-to-timber connections are as illustrated in Tables 4.1 and 4.2. Since the EYM is purposely to be applied on the mortise and tenon and timber-to-timber joints using NDS (2005), thus only the double shear timber-to-timber connections as depicted in NDS (2005) are discussed in this chapter. Double shear joint using the EYM covers the timber-to-timber, timber-to-panel and steel-to-panel connections.

where

$$k_1 = \frac{\sqrt{R_e + 2R_e^2\left(1 + R_t + R_t^2\right) + R_1^2 R_e^3} - R_e(1 + R_t)}{(1 + R_e)}$$

$$k_2 = -1 + \sqrt{2(1 + R_e) + \frac{2F_{yb}(1 + 2R_e)D^2}{3F_{em}l_m^2}}$$

$$k_3 = -1 + \sqrt{\frac{2(1 + R_c)}{R} + \frac{2F_{pb}(2 + R_e)D^2}{3F_{cm}l_s^2}}$$

**Table 4.1** EYM for single shear timber-to-timber equations according to NDS (2005)

| Failure mode | Characteristic load-carrying capacity, Z | Failure mode |
|---|---|---|
| | $\frac{Dl_m F_{em}}{4K_\theta}$ | $I_m$ |
| | $\frac{Dl_s F_{em}}{4K_\theta}$ | $I_s$ |
| | $\frac{k_1 Dl_s F_{es}}{3.6K_\theta}$ | II |
| | $\frac{k_2 Dl_m F_{em}}{3.2(1+2R_e)K_\theta}$ | $III_m$ |
| | $\frac{k_3 Dl_s F_{em}}{3.2(2+R_e)K_\theta}$ | $III_s$ |
| | $\frac{D^2}{3.2K_\theta}\sqrt{\frac{2F_{em}F_{yb}}{3(1+R_e)}}$ | IV |

**Table 4.2** EYM for double shear timber-to-timber equations according to NDS (2005)

| Failure mode | Characteristic load-carrying capacity, Z | Failure mode |
|---|---|---|
| | $\dfrac{Dl_m F_{em}}{4K_\theta}$ | $I_m$ |
| | $\dfrac{2Dl_s F_{es}}{4K_\theta}$ | $I_s$ |
| | $\dfrac{2k_g Dl_g F_{em}}{3.2(2+R_e)K_\theta}$ | $III_s$ |
| | $\dfrac{2D^2}{3.2K_\theta}\sqrt{\dfrac{2F_{em}F_{yb}}{3(1+R_e)}}$ | IV |

$$R_e = \frac{F_{em}}{F_a}$$

$$R_t = \frac{l_m}{l_s}$$

$F_{em}$ = dowel-bearing strength of the main member, psi.

$F_{es}$ = dowel-bearing strength of the side member, psi.

$Z$ = reference design value for fastener in double shear ($Z$ taken as the smallest value from four yield limit equation) lbf.

$$K_\theta = 1 + \frac{\theta}{360}$$

$D$ = fastener diameter, in.

$l_m$ = thickness of the timber middle member, in.

$l_s$ = smaller of the thickness of the timber side member or the penetration depth, in.

$F_{em}$ = dowel-bearing strength of main (centre) member, psi.

$$\begin{cases} F_{ell} \text{ for load parallel to the grain} \\ F_{e\perp} \text{ for load perpendicular to the grain} \end{cases}$$

$F_{e\theta}$ - for load at angle to the grain $\theta$ (see Hankinson formula).

$F_{es}$ = dowel-bearing strength of side member, psi.

$$\begin{cases} F_{ell} \text{ fel load parallel to the grain} \\ F_{e\perp} \text{ fel load perpendicular to the grain} \end{cases}$$

$F_{e\theta}$ for load at angle to the grain $\theta$ (see Hankinson formula).

$F_{yb}$ = bending yield strength of fastener, psi.

$\theta$ is the maximum angle of load to the grain ($0 \leq \theta \leq 90°$) for any member in connection.

Dowel-bearing strength of either the main member or the side member and at an angle of load to grain is given by the Hankinson formula (Equation 4.1):

$$F_{e\theta} = \frac{F_{e//}F_{e\perp}}{F_{e/l}\sin^2\theta + F_{e\perp}\cos^2\theta} \tag{4.1}$$

where

- Yield Mode I = Wood crushing in either the main member or side members. Dowel stiffness is greater than wood strength.
- Yield Mode II = Wood crushing of both main and side member. Dowel stiffness is greater than wood strength.
- Yield Mode III$_m$ = Dowel yield in bending at one plastic hinge point per shear plane and associated wood crushing of main member.
- Yield Mode III$_s$ = Dowel yield in bending at one plastic hinge point per shear plane and associated wood crushing of side members.
- Yield Mode IV = Dowel yield in bending at two plastic hinge points per shear plane and associated wood crushing.

EYM presumes that joints connected with metal dowels will fail in a ductile failure mode (Smith et al. 2001). Mode II and Mode III$_m$ are limited to connections loaded in single shear and are not discussed further in this study. However, four basics model analytically identified by Johansen's in single shear are as depicted by Heine (2001) and shown in Fig. 4.5 for further reference.

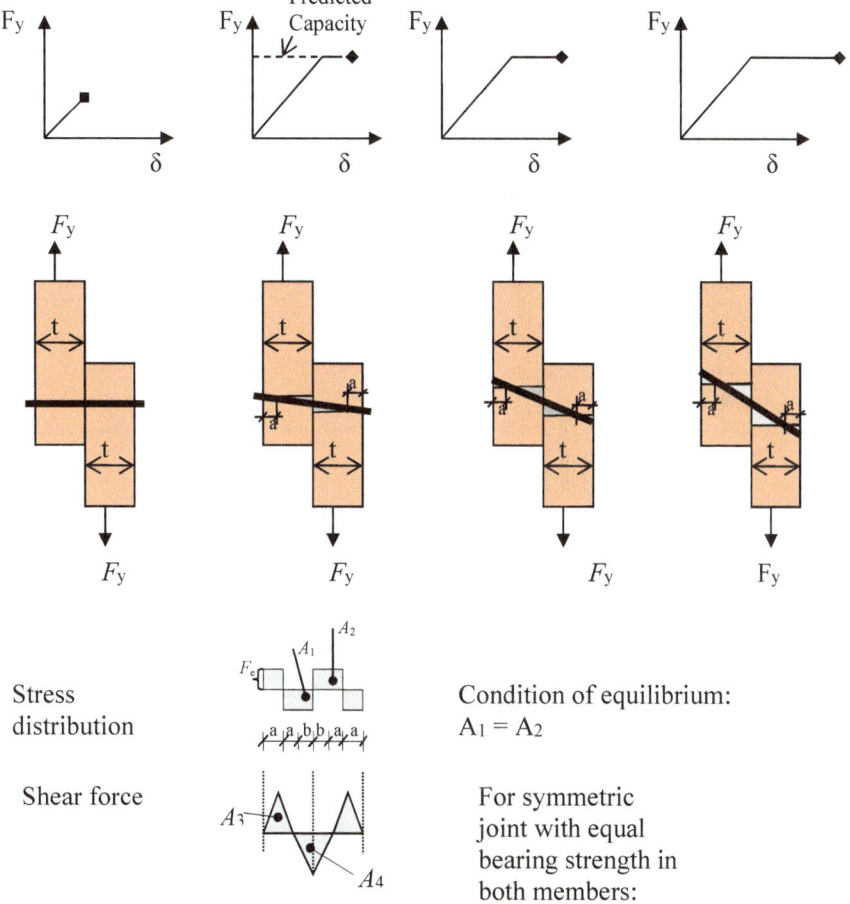

**Fig. 4.5** Johansen's model as adopted from Heine (2001). At capacity, reaction force underneath the dowel is assumed uniformly distributed along the dowel

# References

P. Aune, M. Patton-Mallory, Lateral load-bearing capacity of nailed joints based on the yield theory, in *Forest Products Laboratory Research Paper* (1986)

D.E. Finkenbinder, *An Experimental Investigation of Structural Composite Lumber Loaded by a Dowel in Perpendicular to Grain Orientation at Yield and Capacity.* Master Science Thesis. (Faculty of the Virginia Polytechnic Institute and State University, Virginia, 2007)

K. Harvey, *Improved Timber Connections Using Bonded-in GFRP Rods.* Ph.D. Thesis. (University of Bath, United Kingdom, 2003)

C. Heine, *Simulated Response of Degrading Hysteretic Joints with Slack Behaviour.* Published Ph.D. Thesis (Virginia Polytechnic Institute and State University, Blacksburg, Virginia, 2001)

K.W. Johansen, Theory of timber connection. Int. Assoc. Bridge Struct. Eng. **9**, 249–262 (1949)

M.Z. Jumaat, F. Mohd Razali, A.H. Abdul Rahim, *Development of Limit State Design Method for Malaysian Bolted Timber Joints.* 10th World Conference on Timber Engineering (WCTE). Miyazaki. (Japan, 2008)

H.J. Larsen, The yield load of bolted and nailed connections, in *Proceedings of International Union of Forestry Research Organization, Division V Conference* (1973), pp. 646–655

W.M.C. McKenzie, B. Zhang, *Book of Design of Structural Timber to Eurocode 5*, 2nd edn. (Palgrave Macmillan, 2007)

T.E. McLain, S. Thangjitham, Bolted wood-joint yield model. ASCE J. Struct. Eng. **109**(8), 1820–1835 (1983)

J.F. Miller, R.J. Schmidt, W.M. Bulleit, A new yield model for wood dowel connections. J. Struct. Eng. (2010). https://doi.org/10.1061/(ASCE)ST.1943-541X.0000224

J.F. Miller, *Capacity of Wood dowelled Mortise and Tenon Joints*, Master Science Thesis (Department of Civil and Architectural Engineering, University of Wyoming, Laramie Wyoming, 2004)

NDS, *National Design Specification for Wood Construction American Forest and Paper Association (AFPA)* (Washington, D.C., 2005)

K. Sawata, M. Yasamura, Determination of embedding strength of wood for dowel type fasteners. J. Wood Sci. **48**, 138–146 (2002)

R.J. Schmidt, Timber wood dowels consideration for mortise and Tenon joint design. Wood Des. Focus **14**(3), 44–47 (2006)

R.J. Schmidt, E.D. Daniels, *Design Considerations for Mortise and Tenon Connections.* Report for timber framers guild. (Becket, MA, 1999)

R.J. Schmidt, R.B. MacKay, *Timber Frame Tension Joinery.* Report for timber framers guild. (Becket, MA, 1997)

J.D. Shanks, *Developing Rational Design Guidelines for Traditional Joints in Oak Frame Construction.* Ph.D. Thesis. (University of Bath, 2005)

J.V. Smart, *Capacity Resistance and Performance of Single-shear bolted and Nailed Connections: An Experimental Investigation.* Master Science Thesis (Virginia Polytechnic Institute and State University, Blacksburg, Virginia, 2002)

I. Smith, S.T. Craft, P. Quenneville, Design capacities of joints with laterally loaded nails. Can. J. Civil Eng. **28**, 282–290 (2001)

I. Smith, G. Foliente, M. Nguyen, M. Syme, Capacities of dowel-type fastener joints in Australian pine. ASCE. 17(16), 664 (2005). 10.1061.0899-1561

L.A. Soltis, T.L. Wilkinson, *Bolted-Connection Design* (General Technical Report No. FPL-GTR-54). (Forest Service: Department of Agriculture, United States, 1987)

L.A. Soltis, T.L. Wilkinson, United States adaptation of European yield model to large-diameter dowel fastener specification, in *Proceedings of the 1991 International Timber Engineering Conference: 1991 September 2–5; London*, Vol. 3. (TRADA, London, 1991), pp. 3.43–3.49

Wilkinson, Rowlands, *Western Woods Use Book*, 2nd edn. (Western Wood Product Association, Portland, Oregon, 1979)

T.L. Wilkinson, *Bolted Connection Strength and Bolt Hole Size.* (United States Department of Agriculture: Forest Product Laboratory, 1993), FPL-RP-524

# Chapter 5
# EYM Modification for Wood Dowel Applications

## 5.1 Introduction

In order to apply the EYM equations for the double shear dowelled wood joints to joints that use wooden dowel, the dowel-bearing strength of wood using wood dowels is required (Church and Tew 1997; TFEC 1-2007 2007, TFEC 1-2010 2010). The bending yield strength of the dowel is also needed. The EYM equations may be adapted on all wood joints when the dowel-bearing strength and bending of the wood dowels are known (Church and Tew 1997).

Based on this specific reason, few series of studies concerned with the dowel-bearing strength of dowels were reported. After a pioneered research of wood dowel published by Brungraber (1985), a study on wood dowel connections has been continued at University of Wyoming by a group of researchers led by Schmidt. Most research findings by Schmidt and his group of researchers (Schmidt and MacKay 1997; Schmidt and Daniels 1999; Miller 2004) have contributed to develop TFEC 1-2007 (2007); TFEC 1-2010 (2010), standard as supplement (provisions of the design standard) for timber frame structures for wood construction (ANSI/AFPA NDS 2005). Schmidt and MacKay (1997) modified the EYM into the use of wood dowel specifically for mortise and Tenon joint. Their research discovered additional dowel failure modes (Fig. 5.1) named as mode $III_s$' that were not covered by the existing yield models for connection design.

MacKay's work was then continued by Schmidt and Daniels (1999). Daniels figured out two main types of failure, ductile wood dowel-bearing and brittle connection failure and proposed three new joint failure modes as shown in Fig. 5.2, modes $I_m$ and $I_s$ are the existing EYM. The proposed $I_d$, $III_m$ and $V_d$ are for wood dowel-bearing failure, wood dowel bending failure or shear failure and a shear failure with fractures due to bending near ultimate loads, respectively. Mode $III_s$ as proposed by Schmidt and Mackay (1997) has been named as mode $III_m$ by Schmidt and Daniels (1999).

This work was then continued by Miller (2004) who was also guided by Schmidt. Miller (2004) quantified the shear capacity of wood dowels in a mortise and Tenon

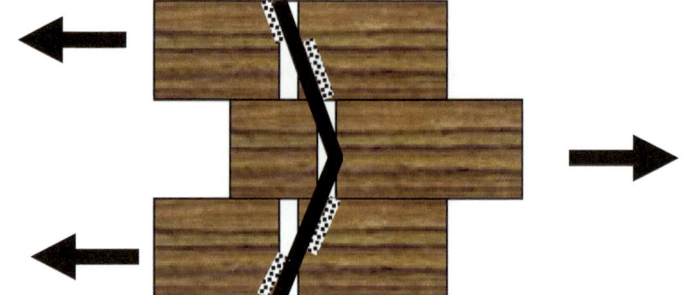

**Fig. 5.1** Type of failure mode III$_s$ as proposed by Schmidt and Mackay (1997)

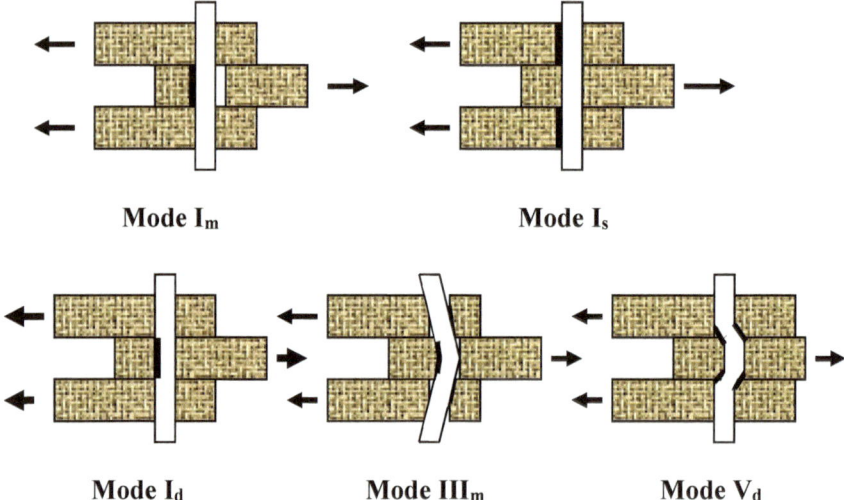

**Fig. 5.2** Joint failure modes as proposed by Schmidt and Daniels (1999)

joint by both physical testing of full-scale specimens as well as modelling their micro-scopic behaviour by the finite element method. Miller (2004) supported Schmidt and Daniels (1999) studies by developing a correlation between shear strength and the specific gravity of the frame materials by testing various species of timber used in both the frame members as well as wood dowels. This correlation was then used to develop a design method for mortise and Tenon. Similar mode as Mode V$_d$ intro-duced by Schmidt and Daniels (1999) is named as Mode V by Miller et al. (2010) and further introduced Eq. (5.2). This Mode V yielding is the fifth failure mode out of four that is stipulated in NDS (2005). Mode V yielding is an effective cross-grain dowel failure which has been observed in physical testing as well as numerical modelling (Fig. 5.3).

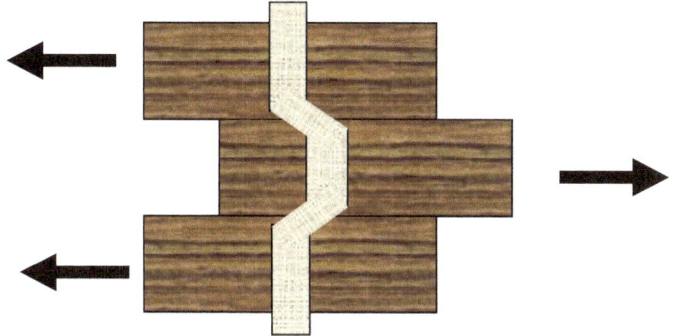

**Fig. 5.3** Mode V of yield mode of failure for doweled connections as proposed by Miller et al. (2010)

$$F_{vy} = 4850G_{dowel}G_{base}^{3/4} \tag{5.2}$$

where $F_{vy}$ is the average Mode V yield shear stress in psi. at the 5% diameter offset yield load, $G_{dowel}$ is the specific gravity of the wood dowel material and $G_{base}$ is the specific gravity of the base (timber) material. This equation was then modified to Eq. (5.3) to consider the reduction term in order to provide a value for Mode V stress at appropriate design load levels.

$$F_v = 1410G_{dowel}G_{Base}^{3/4} \tag{5.3}$$

where $F_v$ is the design yield stress in psi.

A design procedure has also been established for this Mode V yielding by these researchers by being calibrated to the level of performance (reliability) expected from the other yield modes. A regression equation to relate the effective cross-grain yield capacity of wood dowels to specific gravity is also developed for use with the Mode V design equation. Schmidt and Daniels (1999) also claimed that their reliability analysis indicates that the proposed Mode V design equation can be used in conjunction with existing NDS (2005) yield equations for design of wood doweled connections. However, this Mode V design capacity is applicable only for connections in which the dowel is in double shear.

Jumaat et al. (2008) tested double shear joint for Malaysian tropical timber made of Kempas, Mengkulang and Pulai and strengthened using 0.47 in. Diameter steel bolts. They indicated that the failure mode of Kempas falls in range III or IV and EYM (EC 5 2008) has under estimated the ultimate strength for bolted joints up to 74% accordingly.

Above all, in relations of experimental to the EYM equations, Soltis and Wilkinson (1991) stated that the variables of the parameter of interest such as species, moisture content level, types of sides plates (wood or steel), l/d ratio, bolt diameter and grain direction are considered adequate to the EYM when the predictions were within approximately 10% for parallel to the grain and 20% for perpendicular to the grain direction.

## 5.2  Dowel Bending Strength

Connection yield strength is influenced by two parameters that are dowel-bearing strength and dowel bending strength. These parameters are directly related to the capacity and yield mode of the timber joints. The idea of using the full plastic bending capacity to determine joint strength, which is still being used today, was introduced by Meyer (1957). Illustration of Mayer's concept is shown in Fig. 5.4.

Elastic bending moment:

$$M_{el} = \sigma_{max,elastic} \cdot \frac{\pi \cdot d^3}{32} \tag{5.3}$$

Plastic bending moment:

$$M_{pl} = \sigma_{max,plastic} \frac{2 \cdot \pi \cdot r^2}{2} \cdot \frac{4 \cdot r}{3 \cdot \pi} = \sigma_{max} \cdot \frac{d^3}{6} \tag{5.4}$$

where

$M_{max}$   Maximum bending moment, Nmm.

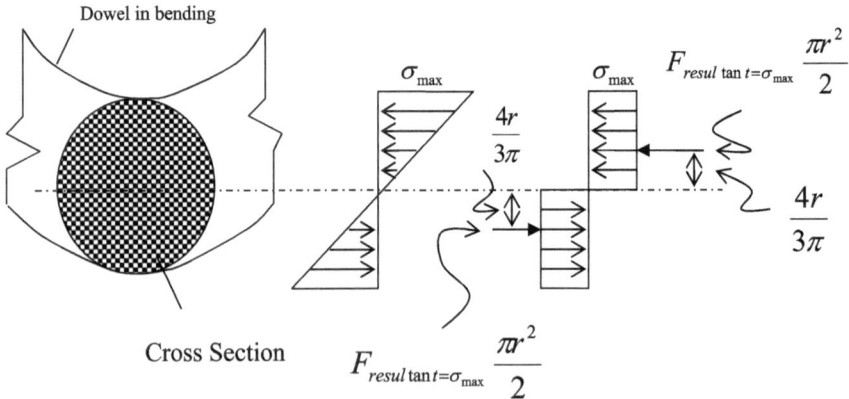

**Fig. 5.4** Cross-section of elastic and plastic bending capacity of the dowel yield bending moment (adopted from Heine 2001)

$\sigma_{max}$    Bending stress or bending yield strength, N/mm$^2$
$d$      Dowel diameter, mm
$r$      $d/2$, mm.
$F$     Force resultant used to derive maximum moment,N

$$\sigma_{max} = \frac{6M_{max}}{d^3} \tag{5.5}$$

Equation (5.4) is used to measure the elastic bending moment, and Eq. (5.5) is used to measure the plastic bending moment theoretically. The bending moment in dowel is determined experimentally. The dowel is bent in a three-point loading.

## 5.3 Dowel-bearing Strength

The dowel-bearing behaviour is defined as the load-deformation behaviour of wood or wood-based products laterally loaded by a fastener where the fastener does not bend during loading (ASTM 5764-9a 2007). Dowel-bearing strength is one of the properties of wood that affects the reference design value capacity (Z) of a nail, bolt or lag bolt. Few literatures termed this property as embedment strength. Dowel-bearing strength is an important parameter to determine EYM.

Since its applications are necessary in predicting load-carrying capacity for timber joints using EYM developed by both NDS (2005) and EC 5 (2008), thus it is mostly expanded in USA and Europe. One of the basic experimental approaches to determine the dowel-bearing strength was started by Wilkinson (1991). During the research work, the author claimed that no standard test exists for evaluating dowel-bearing strength (Wilkinson 1991).

Nevertheless, Wilkinson (1991) also supported the findings by pioneer researchers such as Soltis and Wilkinson (1987) and Whale and Smith (1989). Nowadays, standard to determine the dowel-bearing strength is applicable in ASTM D 5764 (2007) and BS EN383 (2007). Dowel-bearing strength for structural frame in TFEC 1-2007 (2007); TFEC1-2010 (2010) is also based on ASTM 5764; however, TFEC 1-2007 (2007); TFEC1-2010 (2010) does account for the combined response of the timber and the wood dowel fasteners used in connection.

Two types of tests methods commonly used to evaluate the dowel-bearing strength are as specified in NDS (2005)—American Society for Testing Materials (ASTM)–D5764 and EC 5 (2008)—(BS EN 383 2007). ASTM is based on full-hole or the half-hole test while Eurocode 5 is using the full-hole test. According to Awaludin et al. (2007), in half-hole test method; the dowel uniformly loaded along its length, producing a uniform stress distribution through the projected bearing area (Fig. 5.5). All ends of dowels are free to rotate, bearing stress under the dowel is considered to be uniformly distributed and the dowels fit tightly as possible in the hole. Due

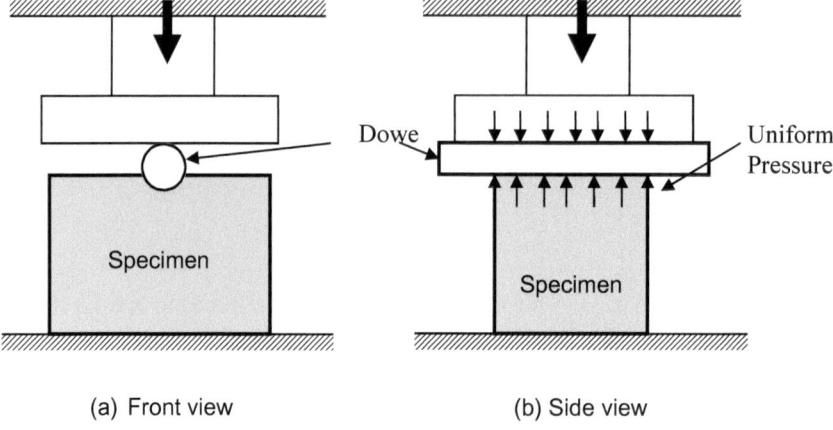

(a) Front view                         (b) Side view

**Fig. 5.5**  Schematic of half-hole test **a** front view and **b** side view

to the basic concept of EYM which ignores friction between the joint members, consequently all friction between the members is not encountered for.

Nonetheless, Heine (2001) claimed that the problem of this method is at lower fastener aspect ratios the wood beneath the fastener tends to split. In the full-hole test, the load applied only at both ends of the dowel. This uneven application of load at both ends might incline the dowel axis or induce some bending in the dowel and produces inconsistent pressure on the specimen (Fig. 5.6).

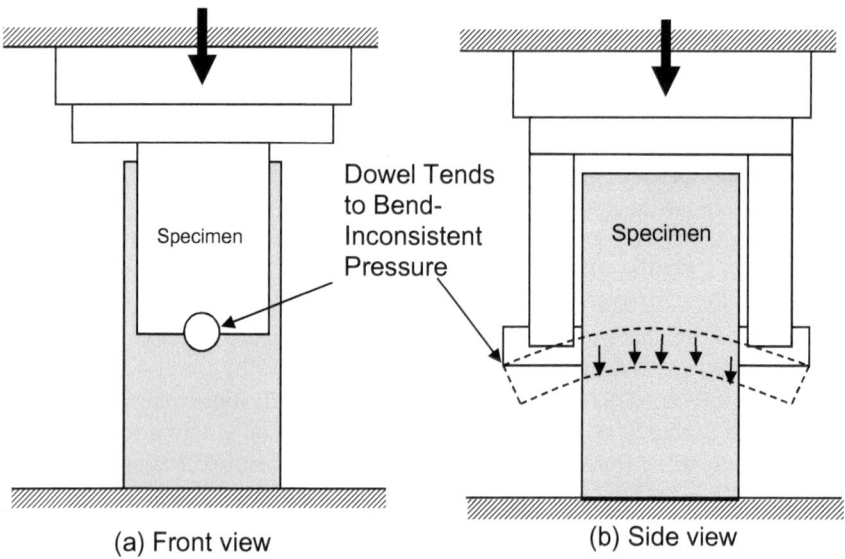

(a) Front view                         (b) Side view

**Fig. 5.6**  Schematic of full-hole test **a** front view and **b** side view

Awaludin et al. (2007) reported that the estimated dowel-bearing strength given by the half-hole test produced higher than that given by the full-hole method or the experimental results. Most researchers such as Wilkinson (1991), Rammer (1999) and Rammer and Winistorfer (2001), Schmidt (2006) and Sauvat et al. (2008) applied the ASTM–D5764 half-hole method in their study. In view of previous studies, the dowel-bearing strength tests in this study were performed in accordance with (ASTM)–D5764 that is using the half-hole test method. This method allows full exposure of the specimens during testing. Thus, detail observations on the specimens during the test such as appearance of cracks or any failure pattern shall be observed. This failure pattern cannot be observed if the full-hole test is used.

Many previous researches relate the dowel-bearing strength to the fastener or the wood characteristics. The fastener's characteristics are such as dowel diameter and dowel type (nails, bolts or screws). Dowel-bearing strength in relations to the wood characteristic is such as moisture content, grain directions, density; specific gravity and timber species. Dowel-bearing strength was also being studied for the engineered wood materials such as glue laminated, laminated veneer lumber and plastic wood composite. (McLain and Thangjitham 1993; Rammer and Winistorfer 2001; Sauvat et al. 2008; Rammer 1999; Jumaat et al. 2006, 2008; Wilkinson 1991; Rammer 1999; Church and Tew 1997; Harada et al. 2000; Hwang and Komatsu 2002; Balma 1999).

As stated in ASTM 5764-97a (2007), a dowel-bearing strength average compressive stress at 5% offset load in a piece of timber or wood-based sheet product under the action of a stiff linear fastener. The fastener mean in this standard is the dowel type fasteners such as bolt, nail or dowel with plain or patterned surfaces. However, as the standard only mentioned that the fastener is not supposed to bend, in this case understandable as steel, it is therefore clearly understandable that this standard is purposely not for other brittle and flexible type of fasteners such as GFRP or wood dowel.

Church and Tew (1997) have introduced a different method in determining the dowel-bearing strength when embedded with wood dowel. They have determined the dowel-bearing strength with an additional experimental method to the ASTM-D5764 and the BS EN 383 (2007) (Fig. 5.7). Instead of pressing the steel dowel with the steel block, Church and Tew have compressed the wood dowel using the wood block.

Their work has been cited by Sandberg et al. (2000) by stated that the nature of Church and Tew's method was by limiting the deformation by both elements; the wood dowel and also the wood-base material (Sandberg et al. 2000). Since the results of Church and Tew's showed a significant reduction of 50% compared to the steel dowel value, thus Sandberg et al. have used different methods in their study with an attempt to separate the wood dowel and wood-base information.

Schmidt and Mackay (1997); Schmidt and Daniels (1999); Miller (2004); Miller et al. (2010) reported about the dowel-bearing strength of wood embedded with wood dowel. Schmidt and Mackay (1997) determined the dowel-bearing strength by directly compressing the wood-base material onto the wood dowel (Fig. 5.8). The outcomes of these tests were then directly applied in the yield model equations. However, no conclusions were made from the produced data regarding the

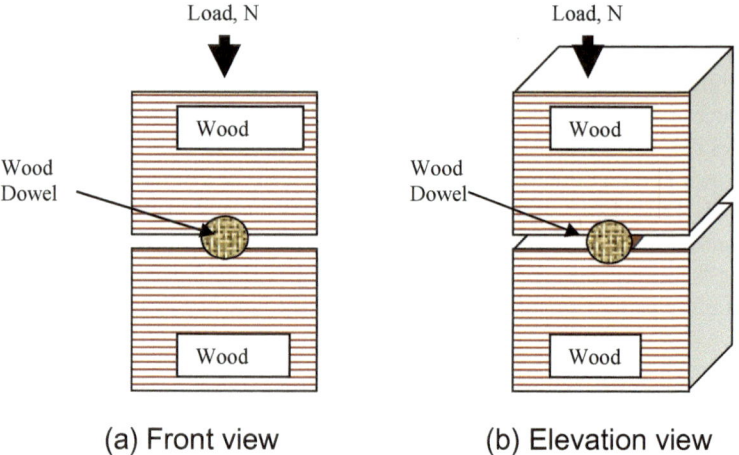

**Fig. 5.7** Method proposed by Church and Tew (1997)

relationship between the wood dowel orientation and the dowel-bearing strength of the material. Using this method, there is also no clear correlations were seen between the yield stresses and the material properties (Schmidt and Mackay 1997).

Another approach by Schmidt and Daniels (1999) to determine the dowel-bearing strength of wood-base material is by introducing the 'Dowel-bearing Spring Theory'. This theory works by combining the load-deformations curve for the base material and wood dowel. The yield strength of the combination is found by applying the 5% offset method to the combined load–deflection curve. Figure 5.9 shows the illustration of this dowel-bearing spring theory.

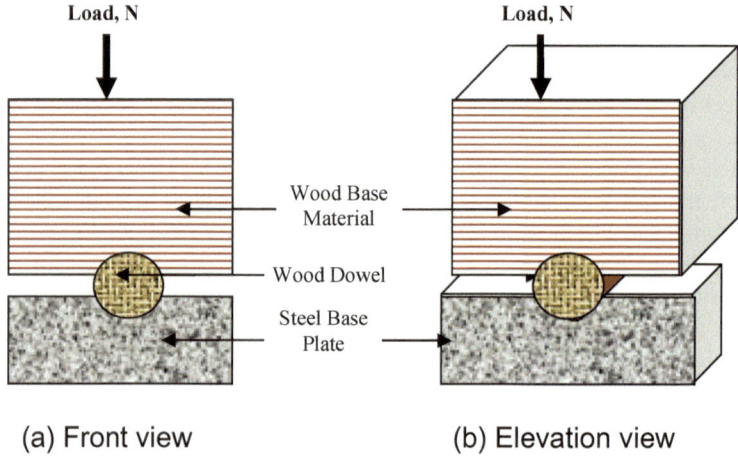

**Fig. 5.8** Method proposed by Schmidt and MacKay (1997)

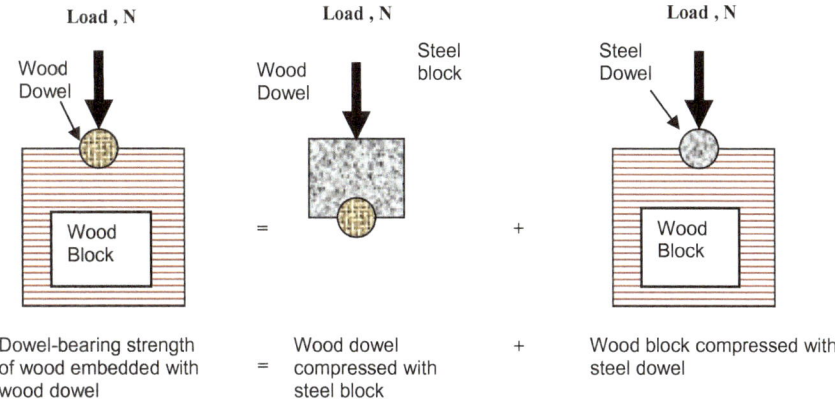

**Fig. 5.9**  Method proposed by Schmidt and Daniels (1999)

This theory is introduced based on the reason that this method can reduce the number of tests to determine the dowel-bearing compressed by wood dowels since the existing data of wood based compressed with steel dowel is widely available for varies species (Schmidt and Daniels 1999). Thus due to the limited existing data, cost and time constrained, the same method as approached by Schmidt and Daniels (1999) for the dowel-bearing strength compressed by wood dowel has not being applied in this research. Therefore, the basic direct bearing method according to the existing standard as stipulated in ASTM-D5764 was made as references in this study.

Though the bearings strength can be easily evaluated from the empirical equations given by the previous studies and current standards, this dowel-bearing strength was developed from softwood species (Rammer 1999; Awaludin et al. 2007; Jumaat et al. 2006, 2008). As claimed by Awaludin et al. (2007), the calculated dowel-bearing strength might be questionable because the equations were developed mainly from the test data of softwood species. Thus, further investigations of dowel-bearing strength for the tropical species are very much needed in order to specifically verify the applicability of the existing standards to the use of tropical species.

Existing data of dowel-bearing strength tested on tropical timber is very much limited. Pioneers in determining the dowel-bearing strength of wood base in Malaysia was started by Jumaat and his group of researchers in 2006 (Jumaat et al. 2006). Their intention was to compile the data of local timber since no study was previously conducted locally which emphasises in this area. Three species to represent three strength groups of Malaysian timber were selected.

Kempas (*Koompassia malaccensis*) and Mengkulang (*Heritiera spp.*) to signify medium hardwood and Pulai (*Alstonia spp.*) selected from light hardwood species. Tests procedures were done according to BS 383 (1993). All tests were done with the length of the dowel parallel to the grain of the base material and compressed with 0.23 in., 0.32 in. and 0.47 in. diameter dowels. Results of dowel-bearing strength were observed in relations to density, and it was found that all species were directly influenced by the density. From the relations of dowel-bearing strength to species;

dowel diameters and density, Jumaat et al. (2006) produced a theoretical formula Eq. (5.6). This equation claimed to be 2 to 20% accuracy with the experimental.

$$F_{e//} = 0.103(1 - 0.013d)\rho \tag{5.6}$$

where

$F_{e//}$   Dowel-bearing strength in perpendicular grain direction.
$\rho$      Density in kg/m$^3$ and.
$d$       is the nail diameter in mm.

Further research in dowel-bearing strength was done in 2008 by the same group of researchers (Jumaat et al. 2008). Five other Malaysian species from different joint group according to MS 544 were tested. The species are Balau (*Shorea Spp.*) from group J1, Kempas (Koompassia Malaccensis) from group J2, Mempening (*Quercus Spp.*) from group J3, Mengkulang (*Heritiera Spp.*) from group J4 and Pulai (*Alstonia Spp.*) from group J5. All specimens were compressed with the length of the dowels parallel to the grain of the base material with 0.31 in., 0.39in. and 0.47 in. diameter dowel.

From their observations, it was concluded that density is a major factor that influences the dowel-bearing strength characteristics. Dowel-bearing strength was also found to be almost constant regardless of dowel diameter except for the timber with greater densities. The dowel-bearing strength decreased slightly as the dowel diameter increased. This study was also found similar to Hilson et al. (1987), Rammer (1999) and Sawata and Yusumura (2002). As an outcome of their study; another dowel-bearing strength theoretical equation in relations with density was developed Eq. (5.7). This equation is claimed to be able to predict the dowel-bearing strength of hardwood species by 0.5% to 7% accuracy. Since the equation Eq. (5.7) is more current than the other equation produced by the same researchers, therefore this study only refers to this equation for analysis and comparison.

$$F_{e//} = 0.0955(1 - 0.02d)\rho \tag{5.7}$$

where

$F_{e//}$   Dowel-bearing strength in parallel grain direction.
$\rho$      Density in kg/m$^3$ and.
$d$       Nail diameter in mm.

Another record on dowel-bearing strength of Balau species with a limited number of tests was published by Awaludin et al. (2007). Only Awaludin et al. (2007) tested on the dowel-bearing strength of hardwood species compressed on different angle which include the parallel and the perpendicular to the grain. They commented that the average dowel-bearing strength parallel to the grain was 7.25% lower that the prediction given in EC 5 (2008). The dowel-bearing strength perpendicular to the grain evaluated based on bearing load at initial cracking was substantially different from any predictions given by previous studies or design standard.

The most recent theoretical dowel-bearing strength formula, that are Eq. (5.8) and (5.9) were proposed by Miller et al. (2010). These equations were developed specifically for dowel-bearing strength embedded with wood dowel.

$$F_{e\perp} = 4900 G_{dowel}\sqrt{G_{base}} \quad \text{unit in psi} \tag{5.8}$$

$$F_{e//} = 4770 G_{dowel}^{1.32} \quad \text{unit in psi} \tag{5.9}$$

where

$F_{e\perp}$    dowel-bearing strength at yield of the base timber when loaded perpendicular to the grain, psi.

$F_{e//}$    dowel-bearing strength at yield of the base timber when loaded parallel to the grain, psi.

In order to apply these equations, the specific gravity of the dowel, $G_{dowel}$ needs to be considered. This parameter is only significant to the dowel-bearing strength of wood base which compressed using wood dowel. Nonetheless, this theory was developed based on limited specific gravity of wood. As specified by Miller et al. (2010), the specific gravity of the dowel must be in the range of 0.57 to 0.75 and $G_{base}$ must be within 0.35 to 0.70. Miller et al. (2010) also claimed that further tests are needed before these equations are applied to other range of specific gravity.

However, Miller et al. (2010) on the other hand did not specifically note either these specific gravities are the value at test, oven dry or at specific moisture content. Nonetheless, the specific gravity, $G_{base}$ value at tests for this finding is within 0.71 to 0.98 and specific gravity, $G_{base}$ at oven dry is within of 0.62 to 0.87. Since the specific gravity of wood base (Kempas and Kapur) is tend to be in a higher group than the formula requirement, thus no further discussion is made between dowel-bearing strength resulted from the observation to the theoretical equation proposed by Miller et al. (2010).

# References

American Forest and Paper Association, *Commentary on the National Design Specification for Wood Construction*. AF&PA. (Washington, D.C., 2005)

American Society for Testing and Materials (ASTM) D 5764, *Standard Test Method for Evaluating Dowel-Bearing Strength of Wood and Wood-Based Products*. (Washington. D.C., (2007)

A. Awaludin, W. Smittakorn, T. Hirai, Bearing properties of *Shorea Obtusa* beneath a laterally loaded bolt. J. Wood Sci. **53**(3), 204–210 (2007)

D.A. Balma, *Evaluation of Bolted Connections in Wood Plastic Composites*. Master Science Thesis. (Washington State University, Washington, 1999)

British Standard (BS) EN 383. *Timber Structures: Determination of Embedment Strength and Foundation Values for Dowel Types Fasteners*. (British Standard Institutes, 2007)

R. Brungraber, *Traditional Timber Joinery: A Modern Approach*. Ph.D. Thesis, (University of Stamford, 1985)

J.R. Church, B.W. Tew, Characterization of bearing strength factors in pegged timber connections. J. Struct. Eng. **12**(3), 326–332 (1997)

Eurocode (EC) 5. *Design of Timber Structures*. (BS EN 1995 -1-1-2004 +A1:2008)

M. Harada, T. Hayashi, M. Karube, K. Komatsu, Dowel-bearing properties of glued laminated timber with a drift pin, in *Proceeding of World Conference in Timber Engineering (WCTE), July 31-August 3, 2000*. (Whistler Resort, British Columbia, Canada, 2000)

C. Heine, *Simulated Response of Degrading Hysteretic Joints with Slack Behaviour*. Published Ph.D. Thesis (Virginia Polytechnic Institute and State University, Blacksburg, Virginia, 2001)

B.O. Hilson, L.R.J. Whale, D.J. Pope, I. Smith, Characteristic properties of nailed and bolted joints under short-term lateral load, part 3—analysis and interpretation of embedment test data in terms of density related trends. J. Inst. Wood Sci **11**(2), 65–71 (1987)

K. Hwang, K. Komatsu, Bearing properties of engineered wood products I: effects of dowel diameter and loading direction. Jpn. Wood Res. Soc. **48**, 295–301 (2002)

M.Z. Jumaat, A. Abu Bakar, F. Mohd Razali, A.H. Abdul Rahim, J. Othman, The determination of the embedment strength of Malaysian hardwood, in *9th World Conference on Timber Engineering (WCTE), 6–10 August* (Portland, OR, USA, 2006)

M.Z. Jumaat, F. Mohd Razali, A.H. Abdul Rahim, Development of limit state design method for Malaysian bolted timber joints. 10th World Conference on Timber Engineering (WCTE). (Miyazaki, Japan, 2008)

T.E. McLain, S. Thangjitham, Bolted wood-joint yield model. ASCE J. Struct. Eng. **109**(8), 1820–1835 (1993)

A. Meyer, Translation by Hein, 2001 of: Die Tragfähigkeit von Nagelverbindungen bei Statischer Belastung. Holz Als Roh- Und Werkstoff. **15**(2), 96–109 (1957)

J.F. Miller, R.J. Schmidt, W.M. Bulleit, A new yield model for wood dowel connections. J. Struct. Eng. (2010) https://doi.org/10.1061/(ASCE)ST.1943-541X.0000224

J.F. Miller, *Capacity of Wood dowelled Mortise and Tenon Joints*, Master Science Thesis. (Department of Civil and Architectural Engineering, University of Wyoming, Laramie Wyoming, 2004)

NDS, *National Design Specification for Wood Construction American Forest and Paper Association (AFPA)* (Washington, D.C., 2005)

D.R. Rammer, Parallel-to-grain dowel-bearing strength of two Guatemalan hardwoods. For. Product J. **49**(6), 77–87 (1999)

D.R. Rammer, S.G. Winistofer, Effect of moisture content on nail bearing strength. Wood Fiber Sci.: J. Soc. Wood Sci. Technol. (USA) **33**(1), 126–139 (2001)

L.B. Sandberg, W.M. Bulleit, E.H. Reid, Strength and stiffness of oak wood dowels in traditional timber-frame joints. ASCE, J. Struct. Eng. **126**(6), 21620 (2000)

N. Sauvat, O. Pop, S. Merakeb, Dubois, Effect of moisture content variation on short term dowel-bearing strength, in *Paper Presented at the 10th World Conference on Timber Engineering (WCTE)*. (Miyazaki, Japan, 2008)

K. Sawata, M. Yasamura, Determination of embedding strength of wood for dowel type fasteners. J. Wood Sci. **48**, 138–146 (2002)

R.J. Schmidt, Timber wood dowels consideration for mortise and Tenon joint design. Wood Des. Focus. **14**(3), 44–47 (2006)

R.J. Schmidt, E.D. Daniels, *Design Considerations for Mortise and Tenon Connections*. Report for Timber Framers Guild. (Becket, MA, 1999)

R.J. Schmidt, R.B. MacKay, *Timber Frame Tension Joinery*. Report for Timber Framers Guild. (Becket, MA, 1997)

L.A. Soltis, T.L. Wilkinson, *Bolted-Connection Design* (General Technical Report No. FPL-GTR-54). (United States, Forest Service: Department of Agriculture, 1987)

L.A. Soltis, T.L. Wilkinson, United States adaptation of European yield model to large-diameter dowel fastener specification, in *Proceedings of the 1991 International Timber Engineering Conference: 1991 September 2–5; London*, vol. 3. (London, TRADA, 1991), pp. 3.43–3.49

TFEC 1-2007, *Standard for Design of Timber Frame Structures and Commentary (Standard)*. (Timber Frame Business Council and Timber Frame Engineering Council, (TFEC-TAC), Becket, 2007)

TFEC 1-2010, *Standard for Design of Timber Frame Structures and Commentary (Standard)*. (Timber Frame Business Council and Timber Frame Engineering Council, (TFEC-TAC), Becket, 2010)

L.R.J. Whale, I. Smith, A method for measuring the embedding characteristics of wood and wood based materials. Mater. Struct. **22**, 403–410 (1989)

T.L. Wilkinson, *Dowel-bearing Strength* (Laboratory Report Research Paper No. FPL-RP-505). (Forest Products Laboratory, One Gifford Pinchot Drive, WI 53705-2398, 1991)

# Chapter 6
# Common Related Parameters

## 6.1 Introduction

Common related parameters summarised in this chapter is the compilations of works that explained the basic parameters which relates to dowel-bearing strength. These parameters of concerns are the dowel diameter; grain directions, moisture contents, specific gravity, density of wood and other wood-based materials such as the glued laminated timber laminated veneer lumber and wood composite material (Church and Tew 1997; Rammer 1999; Harada et al. 2000; Hwang and Komatsu 2002; Rammer and Winistorfer 2001; Whale and Smith 1989; Balma 1999). Figure 6.1 shows the example of dowel-bearing strength wood block specimens after test (a) perpendicular to grain and (b) parallel to grain.

Dowel-bearing strength compressed specifically using wood dowel was observed by Church and Tew (1997), Schmidt and MacKay (1997), Schmidt and Daniels (1999), Soltis and Wilkinson 1987, Whale and Smith 1989, Church and Tew 1997, Harada 2000, Heine 2001, Hwang and Komatsu 2002, Miller 2004, Miller et al. (2010).

## 6.2 Influence of Dowel Diameter and Grain Directions

Many reports relate the dowel-bearing strength with the dowel diameter. However, varies parameter of interest that influence the dowel diameter used in these studies made the direct comparison not possible. Previous findings relate the dowel-bearing strength and the dowel diameter using hardwood (Rammer 1999), Jumaat et al. (2006, 2008), softwood (Rammer and Winistorfer 2001), glued laminated wood (Church and Tew 1997; Harada et al 2000) and laminated veneer lumber (Hwang and Komatsu 2002). Each of these findings was tested based on different base materials and different sizes of dowel diameters made the direct comparison is not viable.

R. Hassan et al., *Timber Connections*, SpringerBriefs in Applied Sciences and Technology, https://doi.org/10.1007/978-981-19-2697-6_6

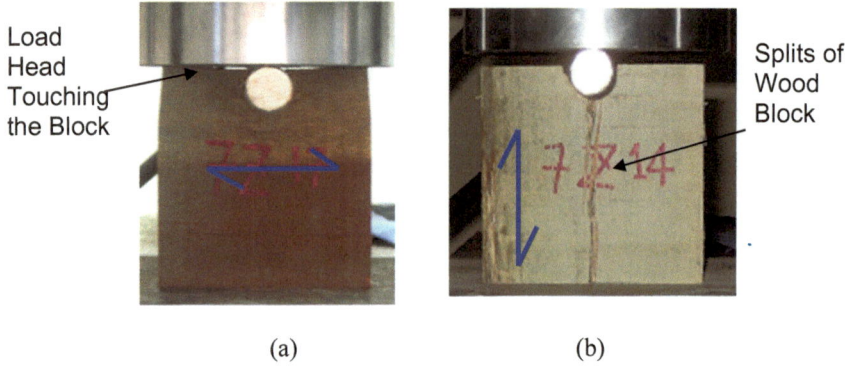

Load
Head
Touching
the Block

Splits of
Wood
Block

(a)                                               (b)

**Fig. 6.1** Example of dowel-bearing strength wood block specimens after test **a** perpendicular to grain and **b** parallel to grain

Other parameters of concerns that defers are the grain direction of the base material and as well as the dowel material. The direction of perpendicular to the grain, parallel to the grain, tangential or radial has significantly influenced their findings. These parameters were also different when they are compared using the different basic properties of the base material such as the specific gravity, moisture content and the density.

The dowel might as well differ in the form of nails (Rammer 1999; Rammer and Winistorfer 2001), bolts (Rammer 1999; Rammer and Winistorfer 2001; Wilkinson 1991), screws (Chui et al. 2006) or the smooth dowel (Harada et al. 2000; Hwang and Komatsu 2002). Thus, little direct comparison can be done and discussed throughout these compilations of work. However, brief explanations related to the influence of the dowel diameter and grain directions are subsequently reported in this chapter. Church and Tew (1997) studied on wood dowel embedded on glued laminated timber made of oak and Douglas fir.

Equations given in NDS 2005 which have been established for timber joints using metal fasteners have been applied in their study. They stated that no dowel-bearing strength values, such as those found in NDS 2005 based on steel fasteners, exist for joints using wooden dowels. Therefore, they have claimed that their work serves as an initial base upon which to develop appropriate design equations for wood dowel timber joints. According to them, the full-scale joint testing can be done to verify the appropriateness of these equations.

Church and Tew (1997) reported their findings based on the influence of different dowel diameters, tangential and radial to the grain according the proportional limit stress and 5% offset stress value. They notified that the effect of varying the wood dowel diameter has very little on the proportional limit stress or the 5% offset stress. Their observations were based on the orientation of the wood dowel that is in the tangential and the radial grain direction.

When the wood dowel loaded in the tangential direction, no significant difference was found between the proportional limit stresses or the 5% offset for 1, 1.13 and 1.35

in. dowels. However, the proportional limit stress was lower for the 1 in. dowels than for the 1.13 and 1.25 in. dowels in the radial direction. Two dowel-bearing strength of Guatemalan hardwood species that is Danto (*Brosimum alicastrum*) and Ramon (*Vatairea lundellii*) were described in relations to the parallel and perpendicular to the grain by Rammer (1999). The author's objective was to determine the dowel-bearing strength of higher specific gravity to augment the data base for determining bolt-and-nail dowel-bearing strength and specific gravity relationships.

As a result, Rammer (1999) indicated that bolt-bearing stiffness is influenced by the diameter of the bolt, but bolt-bearing stiffness and strength are significantly influenced by the diameter of the nail. Comparison of the dowel-bearing strength of a similar sized nail and bolt indicates a statistically significant difference. Finally, Rammer (1999) also concluded that the existing USA and European expressions relating dowel-bearing strength to specific gravity and dowel diameter fail to predict mean values or trends in these high-density timber species.

Smooth shank nails of 0.5 and 0.13 in. diameter bolts were tested on parallel and perpendicular to the grain on Southern pine, Douglas fir-larch and spruce pine-fir by Rammer and Winistorfer (2001). These two diameter bolts were observed in relation of moisture content to the dowel-bearing strength. Rammer and Winistorfer (2001) revealed that the relationship for dowel-bearing strength and moisture content is independent of dowel diameter for the two diameters tested.

Chui et al. (2006) studied on the effects of dowel size on lateral strength of steel to wood screw joints using the single shear steel plate to wood screw. Steel wood screws of shank diameters of 0.17 and 0.22 in.; self-tapping screws of 0.24 in. diameter and steel lag screws of 0.25 in. in diameter were used to encompassed the influence of screw diameter for load parallel to grain or perpendicular to grain. They concluded that the difference between dowel-bearing strength perpendicular to the grain and parallel to the grain increases with screw diameter.

Dowel-bearing properties of glued laminated timber with a drift pin have been reported by Harada et al. (2000) using ASTM 5764, 1997, (half-hole test). The accuracy of present formulas for mechanical joints of glued laminated timbers (GLTs), dowel-bearing tests with drift pin were conducted in their study. The GLTs were composed of mechanically graded Japanese cedar and Japanese larch laminae with uniform modulus of elasticity. Specimens were largely divided into two types parallel and perpendicular to the grain. Five drift pins with diameter of 0.16, 0.31, 0.47 and 0.79 in. were embedded into GLTs.

As a result, they have obtained that 5% offset values were constant regardless of the diameter for the parallel whereas declining tendency with increasing diameter for the perpendicular type. Initial stiffness decreased with increasing diameter for the perpendicular type, but this tendency was not clear for the parallel type.

Effective elastic foundation depth increased with increasing diameter for both types, Hirai's formula commonly used to estimate the relationship between the depth and diameter of fasteners was applied to these results, however, the correlation was low. This discrepancy resulted from the fact that the formula was derived from experiments in which fasteners of small diameters were used.

Effects of dowel diameter and loading direction of laminated veneer lumber (LVL) with two moduli elasticity (MOE; 7.8 and 9.8 GPa) have been studied by Hwang and Komatsu (2002). Tests were conducted according to ASTM D5764 with consideration on the engineered wood product (EWP) anisotropic structures. They claimed that when formulating the new structural design formulas, it is difficult to estimate the initial stiffness (bearing constant) perpendicular to the grain. According to the definition of initial stiffness, the initial stiffness perpendicular to the grain should be estimated based on modulus of elasticity (MOE) at 90°, the MOE perpendicular to the grain. However, this information is limited.

Two per cent (2%) and five per cent (5%) offset were compared in analysing the dowel-bearing strength and trend of variation as the dowel diameter increased. This is because most specimens split before reaching 5% offset when dowel diameter increased. In most of the specimens, the 5% offset seems to be proportional to 2% offset. Results showed that the different in structure of the specimens will cause effects on the results differences.

Hwang and Komatsu (2002) have concluded that the bearing properties of the different EWPs are more depended on their elemental structures. Except in a few cases that none of the four EWPs showed a clear relation between the dowel-bearing strength and dowel diameter, irrespective of loading direction. They found that the initial stiffness of these EWPs decreased with increasing dowel diameter. This statement of behaviour found to be similar to the behaviour on glued laminated timber by Harada et al. (2000). They have also commented that it is necessary to derive the bearing capabilities based on the different equations, taking into consideration the loading direction based on the differences of bearing characteristics shown between the EWPs.

Soltis and Wilkinson (1987) tested joints with nail diameters ranging from 0.1 to 0.24 in. and plain dowels from 0.11 to 0.49 in. diameter and claimed that there is no difference in lateral strength for nail joints loaded parallel or perpendicular to the grain for all sizes. Their study was supported by Whale and Smith (1989) for the small-diameter findings. Whale and Smith (1989) tested for nails to 0.24 in. in diameter and found that the difference between parallel and perpendicular to the grain embedment strengths is small and by extension that joint design strengths can be independent of direction of loading.

Wilkinson (1991) reported that the NDS 2005 (1986) did recognises no difference between parallel and perpendicular loading of small-diameter dowels, such as nails, but does recognise such a difference for large diameter dowels, such as bolts. However, in contradict to the small diameter as reported by Soltis and Wilkinson (1987), Whale and Smith (1989), Wilkinson (1991), Rammer (1991) tested on hardwood species and found that dowel diameter does influence the dowel-bearing strength when compressed on hardwood (Rammer 1999) but independent when in relation to the moisture content of softwood (Rammer and Winistorfer 2001).

Little effects of dowel diameter were found when it was compressed on glue laminated wood base (Church and Tew (1997), but declining when compressed with bigger diameter (Harada 2000). Dowel-bearing strength of LVL shows a similar pattern to glue laminated since Hwang and Komatsu (2002) proved that based on the initial stiffness, the dowel-bearing strength is decreased when the dowel diameter is increased. However, these results opposed the findings by Chui et al. (2006) when it is compressed using wood screw, which concluded that the dowel-bearing strength perpendicular to the grain and parallel to the grain increases with screw diameter.

Gattesco (1998) claims that embedment strength is considerably influenced by loading direction (parallel or perpendicular to grain) not only because of different modulus of elasticity of wood when compressed parallel or perpendicular to grain, but also on account of different failure modes.

Sawata and Yasumura (2002) have applied the EC 5 method, yet analysed both using the 5% offset method according to NDS 2005 and the 0.20 in. displacement as requested by EC 5, 2008. The results show that the dowel-bearing strength parallel to the grain was 0.9 times as large as the compressive strength parallel to the grain regardless the evaluation method. The dowel-bearing strength perpendicular to the grain evaluated by 5% offset method was four times as large as the compressive strength perpendicular to the grain.

When the embedding strength perpendicular to the grain was evaluated by a maximum load up to 0.20 in. displacement according to EC 5, the ratio of embedding strength perpendicular to the grain to the compressive strength perpendicular to the grain decreased as the dowel diameter increased.

Miller (2004) has tested 26 dowel-bearing tests using steel dowels in 1 in. diameter embedded on similar sizes of yellow poplar block half-hole specimens in accordance to ASTM D5464. Thirteen samples were tested in parallel to the grain, and thirteen samples were tested in perpendicular to the grain. The specimens were loaded at a rate of 0.04 inches per minute until the ultimate load was achieved. He reported that for parallel to the grain loading, yellow poplar was 3.5 times stiffer and 2.2 times stronger at yield than for perpendicular to the grain.

## 6.3  Influence of Specific Gravity

Embedment strength for parallel and perpendicular to the grain loading has been related to the bolt diameter and specific gravity by Wilkinson (1991) using empirical equation (Eqs. 6.1 and 6.2). These equations were derived based on the 5% offset load rather than capacity and is applied in the current NDS 2005 after a few modifications. These equations are applied as the comparison to the experimental results in this study.

$$F_{e//} = 11,200G \tag{6.1}$$

$$F_{e\perp} = 6,100G^{1.45}d^{-0.5} \tag{6.2}$$

where

$F_{e\perp}$   dowel-bearing strength in the perpendicular grain direction, psi

$F_{e//}$   dowel-bearing strength in parallel grain direction, psi.
$d$      fastener diameter, in.
$G$      specific gravity based on oven dry weight.


## 6.4  Influence of Moisture Content

Rammer and Winistorfer (2001) reported on the dowel-bearing strength values in relation with moisture content obtained from two groups of softwood specimens. They have suggested that the dowel-bearing strength increases with decreasing moisture content, much like other wood properties. An empirical linear model (Eq. 6.3) is presented relating the two is described as follows with $r^2 = 0.52$.

$$F_e = 489.95 - 1.186m \tag{6.3}$$

where

$F_e$   dowel-bearing strength
$m$    moisture content.


## 6.5  Influence of Density

According to Whale and Smith (1989), the embedding strength for timber based on a large number of embedding tests can be expressed as Eqs. 6.4 and 6.5 depending on the fastener diameter and the timber density for loads parallel to the grain direction:

$$F_{e//} = 0.082\rho d^{-0.3} \text{ N/mm}^2 \quad \text{without predrilled holes and} \tag{6.4}$$

$$F_{e//} = 0.082(1 - 0.01d)\rho_k \text{ N/mm}^2 \quad \text{with predrilled holes} \tag{6.5}$$

where

$F_{e//}$   dowel-bearing strength in perpendicular grain direction
$\rho$      density in kg/m$^3$ and $d$ the nail diameter in mm.

These equations have been applied in EC 5, 2008, but only for the nails up to 0.24 in. diameter.

Dowel-bearing strength formula (Eq. 6.6) for hardwood for the dowel up to 1.18 in. diameter from EC 5, 2008 is similar to bolts supported by the Eqs. 6.7 and 6.8:

$$f_{h,\alpha,k} = \frac{f_{h,0,k}}{k_{90} \sin^2 \alpha + \cos^2 \alpha} \tag{6.6}$$

where

$$k_{90} = 0.90(1 - 0.01d)\rho_k \tag{6.7}$$

$$f_{h,0,k} = 0.082(1 - 0.01d)\rho_k \text{ N/mm}^2 \tag{6.8}$$

where

$f_{h,0,k}$  dowel-bearing strength in perpendicular grain direction
$\rho$  density in kg/m$^3$ and d the dowel diameter in mm.

Influence of density to dowel-bearing strength for hardwood using Malaysian timber species that is Balau, Kempas, Mempening, Mengkulang and Pulai have been studied by Jumaat et al. (2008). They proposed Eqs. 6.7 and 6.8 to estimate dowel-bearing strength in compression parallel to the grain of these species. An average result from 3 numbers of tests for each species was reported. These researchers concluded that the dowel-bearing strength of Malaysian timber species was significantly affected by the density. These values were found to be 0.5 and 7% larger than those that were derived from EC 5, 2008 equation.

# References

D.A. Balma, *Evaluation of Bolted Connections in Wood Plastic Composites*. Master Science Thesis, Washington State University, Washington, 1999

Y.H. Chui, I. Smith, Z. Chen, Influence of fastener size on lateral strength of steel-to-wood screw joints. For. Prod. J. Madison **56**(7/8), 49 (2006)

J.R. Church, B.W. Tew, Characterization of bearing strength factors in pegged timber connections. J. Struct. Eng. **12**(3), 326–332 (1997)

N. Gattesco, Strength and local deformability of wood beneath bolted connectors. ASCE J. Struct. Eng. **124**(2), 195–202 (1998)

M. Harada, T. Hayashi, M. Karube, K. Komatsu, Dowel-bearing properties of glued laminated timber with a drift pin, in *Proceeding of World Conference in Timber Engineering (WCTE)*, July 31–Aug 3, 2000. Whistler Resort, British Columbia, Canada, 2000

C. Heine, *Simulated Response of Degrading Hysteretic Joints with Slack Behaviour*. Published Ph.D. Thesis, Virginia Polytechnic Institute and State University, Blacksburg, Virginia, 2001

K. Hwang, K. Komatsu, Bearing properties of engineered wood products I: effects of dowel diameter and loading direction. Jpn. Wood Res. Soc. **48**, 295–301 (2002)

M.Z. Jumaat, A. Abu Bakar, F. MohdRazali, A.H. Abdul Rahim, J. Othman, The determination of the embedment strength of Malaysian hardwood, *in 9th World Conference on Timber Engineering (WCTE)*, 6–10 Aug, Portland, OR, USA, 2006

M.Z. Jumaat, F. MohdRazali, A.H. Abdul Rahim, Development of limit state design method for Malaysian bolted timber joints, in *10th World Conference on Timber Engineering (WCTE)*. Miyazaki. Japan, 2008

J.F. Miller, *Capacity of Wood dowelled Mortise and Tenon Joints*, Master Science Thesis, Department of Civil and Architectural Engineering, University of Wyoming, Laramie Wyoming, 2004

J.F. Miller, R.J. Schmidt, W.M. Bulleit, A new yield model for wood dowel connections. J. Struct. Eng. (2010). https://doi.org/10.1061/(ASCE)ST.1943-541X.0000224

D.R. Rammer, Parallel-to-grain dowel-bearing strength of two Guatemalan hardwoods. For. Prod. J. **49**(6), 77–87 (1999)

D.R. Rammer, S.G. Winistofer, Effect of moisture content on nail bearing strength. Wood Fiber Sci. J. Soc. Wood Sci. Technol. **33**(1), 126–139 (2001)

K. Sawata, M. Yasamura, Determination of embedding strength of wood for dowel type fasteners. J. Wood Sci. **48**, 138–146 (2002)

R.J. Schmidt, R.B. MacKay, Timber Frame Tension Joinery. Report for Timber Framers Guild. Becket, MA, 1997

R.J. Schmidt, E.D. Daniels, Design Considerations for Mortise and Tenon Connections. Report for Timber Framers Guild. Becket, MA, 1999

L.A. Soltis, T.L. Wilkinson, *Bolted-Connection Design* (General Technical Report No. FPL-GTR-54). United States, Forest Service, Department of Agriculture, 1987

L.R.J. Whale, I. Smith, A method for measuring the embedding characteristics of wood and wood based materials. Mater. Struct. **22**, 403–410 (1989)

T.L. Wilkinson, *Dowel-bearing Strength* (Laboratory Report Research Paper No. FPL-RP-505): Forest Products Laboratory, One Gifford Pinchot Drive, WI 53705-2398, 1991

# Chapter 7
# Factor of Safety

## 7.1 Introduction

Factor of safety is the ratio of the ultimate stress during working to the actual or allowable applied stress in service. The actual applied stress has to be much smaller than the ultimate stress to have a sufficient safety factor (Sarkar 2003). Therefore, Eq. 7.1 is used to predict the factor of safety.

$$\text{FoS} = \frac{\text{Ultimate Stress}}{\text{Allowable/working Stress}} \tag{7.1}$$

The value of the ultimate stress comes from the ultimate load to which a member is subjected to an actual working load that it can withstand. This result may be gathered from experimental or physical tests. The applied stress is the maximum allowable stress that the member is allowed to carry in terms of design capacity during the services of the member. This value may be gathered from the analytical formulas taking the actual values as the guidance. According to Sarkar (2003), the factor of safety depends on many other factors such as the homogeneity of the materials used and the accuracy with which stresses in members can be evaluated.

Sarkar also suggested that for most engineering structures, a factor of safety between 2 and 6 is suitable, while Madsen (1992), Smith et al. (2001) mentioned that the design capacities of joints with laterally loaded nails designed by North American engineers are based on working stress using a factor of safety of 1.3 for members loaded in bending, tensile and compression.

For wood dowelled mortise and tenon joints loaded in tensile, a factor of safety of 3.0 from an average of 3 numbers of specimens in a group is recommended by Kessel and Augustin (1996). The factor of safety recommended by Kessel and Augustin (1996) was based on the ratio of ultimate strength from the experimental to the predicted/allowable load-carrying capacity values. Schmidt and Daniels (1999) recommended a factor of safety of 2.0 for mortise and tenon joints using the ratio of average 5% offset yield to the predicted/allowable load-carrying capacity value.

R. Hassan et al., *Timber Connections*, SpringerBriefs in Applied Sciences and Technology, https://doi.org/10.1007/978-981-19-2697-6_7

**Table 7.1** Allowable joint loads produced by Schmidt and Daniels (1999) using Kessel and Augustin (1996) suggested factors of safety

| | Joint load (lbs) | | | Factor of safety | Recommended loads (lbs) | | |
|---|---|---|---|---|---|---|---|
| | Southern yellow pine (SYP) | Douglas fir (DF) | Red oak (RO) | | Southern yellow pine (SYP) | Douglas fir (DF) | Red oak (RO) |
| Mean ultimate | 6720 | 6500 | 9100 | 3.00 | 2240 | 2170 | 3030 |

**Table 7.2** Recommended factor of safety by Schmidt and Daniels (1999)

| | Joint groups | | |
|---|---|---|---|
| | SYP | DF | RO |
| Using Kessel's recommended FoS (lbs) | 2240 | 22170 | 3030 |
| Experimental data (lbs) | 4210 | 4770 | 5550 |
| FoS for 5% offset strength | 1.9 | 2.2 | 1.8 |

Mean factor of safety = 2.0

Shown in Tables (7.1) and (7.2) are the minimum values produced from Schmidt and Daniels (1999) using procedure outlined by Kessel and Augustin (1996). The allowable capacity for all joint groups was the mean ultimate load divided by a factor of safety of 3.0.

As shown in the table, the factors were averaged to determine a common value for timber frame joint design. The resulting factors of 2.0 found by scientist Schmidt and Daniels (1999) convert the 5% offset load to the recommended design value for short duration load.

Miller (2004) determines the factor of safety for mortise and tenon joints loaded in shear using a ratio of a correlation equation produced at yield load introduced to the existing EYM, NDS 2005 formulas. The factor of safety produced by Miller (2004) for the wood dowelled mortise and tenon joints loaded in shear is 2.2.

Concept of factor of safety in used in relations to the allowable design strength to 5% offset load and the ultimate load as a comparison to prove the reliability of the EYM in predicting the joint connected with GFRP dowel. The similar process was also used for joint connected with wood dowel for comparison purposes.

As a short conclusion, this book covers the explanation about the EYM as the only available yield model being applied in designing timber joints. EYM has been developed for timber joints strengthened with steel fastener, nevertheless few modifications of the EYM were found suggested in some studies for the use of the timber joints strengthened with wood dowels. However, up to this date of writing, no standard is yet produced for the modified EYM specifically for the wood dowels application.

In order to analysed the EYM reliability on other type of fasteners such as the GFRP and wood dowels, the dowel bending strength and dowel-bearing strength of wood-base materials were critically reviewed in this chapter. Many previous works were found contributed to determine on the factors that affected the timber joints load-carrying capacities such as the dowel diameter, grain directions, specific gravity, moisture content and density. However, most of the studies were reported based on the European timber species and very limited information was found reported on tropical timber species.

A factor of safety determined by the previous researches as part of timber design information that relates the ratio of the ultimate stress during working to the actual or allowable applied stress in service were also discussed in this book. The concept of factor of safety was taken as the guideline on the reliability of the EYM equations on predicting the timber joints load-carrying capacity.

# References

M.H. Kessel, R. Augustin, Load behaviour of connections with oak pegs, translation by Schmidt, R., Timber Framing. J. Timber Framers Guild North Am. **39**, 8–11 (1996)

B. Madsen, *Structural Behaviour of Timber* (Timber Engineering Ltd., North Vancouver, B.C., 1992), p. 437. ASBN 0-9696162-001

J.F. Miller, *Capacity of Wood dowelled Mortise and Tenon Joints*, Master Science Thesis, Department of Civil and Architectural Engineering, University of Wyoming, Laramie Wyoming, 2004

B.K. Sarkar, *Strength of Materials* (Tata McGraw-Hill Publications, 2003)

R.J. Schmidt, E.D. Daniels, Design Considerations for Mortise and Tenon Connections. Report for Timber Framers Guild. Becket, MA, 1999

I. Smith, S.T. Craft, P. Quenneville, Design capacities of joints with laterally loaded nails. Can. J. Civ. Eng. **28**, 282–290 (2001)

# Bibliography

American Society for Testing and Materials. Standard Test Method for Determining Bending Yield Moment of Nails. ASTM F 1575-03 (ASTM, Philadelphia, Pa, 2008)

American Society for Testing and Materials. Standard Test Method for Bolted Connections in Wood and Wood-Based Products. ASTM D 5652-95 (ASTM, Philadelphia, Pa, Reapproved 2007)

American Society for Testing and Materials. ASTM D143-94. Standard Methods of Testing Small Clear Specimens of Timber (Washington. D.C., 1999a)

American Society for Testing and Materials. ASTM D4442-07, *Standard Test Method for Direct Moisture Content Measurement of wood and Wood-Base Material* (Washington. D.C.)

BS373, *Methods of Testing Small Clear Specimens of Timber* (British Standards Institution, 1957)

British Standard (BS) EN 5268, *Structural Use of Timber: Part 2: Code of Practice for Permissible Stress Design, Material and Workmanship* (British Standard Institutes, 2002)

E.A.R. Chik, *Malaysian Timbers—Kapur. Malaysia Forest Service. Timber Leaflet* (No. 46, Forest Department of Peninsular Malaysia, Kuala Lumpur, Malaysia, 1988)

J.D. Dolan, S.T. Gutshall, T.E. McClain, *Monotonic and Cyclic Tests to Determine Short-Term Load Duration Performance of Nail and Bolt Connections. Vol. 1: Summary Report* (Virginia Polytechnic Institute and State University Timber Engineering Report No. TE-1994-001, 1995)

C.A. Eckelman, E. Haviarova, H. Akcay, Parallel-to-grain end-load capacity of round mortises in round and rectangular timbers. For. Prod. J. **57**(4), 66–71 (2007)

C.A. Eckelman, E. Haviarova, Load capacity and deflection characteristics of large wooden dowels loaded in double shear. For. Prod. J. **57**(5), 60–64 (2007)

R. Hassan, *Structural Performance of GFRP Dowelled Mortise and Tenon Connections Made of Selected Tropical Species*. Ph.D. Thesis, Faculty of Civil Engineering, Universiti Tenologi MARA. Shah Alam, Selangor, Malaysia, 2011

T. Hirai, Basic properties of mechanical wood-joints II: bearing properties of wood under a bolt. Res. Bull. Coll. Exp. for. Faculty **46**, 967–988 (1989)

A. Jorissen, *Double Shear Timber Connections with Dowel Type Fasteners*. Ph.D. Dissertation, Delft University of Technology, Delft, The Netherlands, December, 1998

H.J. Larsen, S. Thelanderson, *Timber Enginnering* (Wiley, 2003), pp. 316–322. ISBN 0-470-84469-8

T.E. McLain, Connectors and fasteners: research needs and goals, in *Wood Engineering in the 21st Century: Research Needs and Goals* (ASCE, Reston, VA, (1998), pp. 56–69

M.J. Norusis, *SPSS 7.5, Guide to Data Analysis* (Prentice Hall, 1997). ISBN 0-13-656877-7

M. Patton-Mallory, P.J. Pellicane, F.W. Smith, Modeling bolted connections in wood: review. ASCE J. Struct. Eng. **123**(8), 1054–1062 (1997)

© The Editor(s) (if applicable) and The Author(s), under exclusive license to Springer Nature Singapore Pte Ltd. 2023
R. Hassan et al., *Timber Connections*, SpringerBriefs in Applied Sciences and Technology, https://doi.org/10.1007/978-981-19-2697-6

D.J. Pope, B.O. Hilson, Embedment testing for bolts and comparison of the European and American procedures. J. Inst. Wood Sci. **13**(6, Issue 78), 568–571 (1995)

C.Y. Pun, *Structural Timber Joints*. (Malayan Forest Records No. 32) (Forest Research Institute Malaysia. FRIM. Malaysia, 1987)

E.H. Reid, *Behaviour of Wood Pegs in Traditional Timber Frame Connections*. Master Science Thesis, Michigan Technological University, Houghton, Michigan, 1997

A. Reiterer, G. Sinn, S.E. Stanzl-Tschegg, Fracture characteristics of different wood species under mode I loading perpendicular to the grain. J. Mater. Sci. Eng. **A332**, 29–36 (2002)

L.A. Soltis, R.J. Ross, D.F. Windorski, Fibreglass-reinforced bolted wood connections. For. Prod. J. ABI Inform Glob. **48**(9), 63–67 (1998)

L.A. Soltis, Bolted connection research: present and future. Wood Des. Focus **5**(2) (1994)

I. Smith, G. Foliente, M. Nguyen, M. Syme, Capacities of dowel-type fastener joints in Australian pine. ASCE 10.1061(0899-1561) **17**(16), 664 (2005)

D. Tan, S. Ian, Failure in-the-row model for bolted timber connections. J. Struct. Eng. **12**(7), 713–718 (1999)

A.N. Tankut, N. Tankut, The effects of joint forms (shape) and dimensions on the strengths of mortise and tenon joints. Turk. J. Agric. **29**, 493–498 (2004)

G.W. Trayer, The bearing strength of wood under bolts, USDA Washington. Tech. Bull. **332** (1932)

L.R.J. Whale, I. Smith, Mechanical timber joints, in *Research Rep. No. 18/86* (Timber Research and Development Association, High Wycombe, U.K., 1987)

# Index